# ZIML Math Competition Book

## Division H 2017-2018

## Areteem Institute

**ZIML Math Competition Book Division H 2017-18**

Edited by   John Lensmire
David Reynoso
Kevin Wang
Kelly Ren

Copyright © 2018 ARETEEM INSTITUTE
WWW.ARETEEM.ORG

ISBN: 1-944863-29-X
ISBN-13: 978-1-944863-29-6

*First printing, November 2018.*

TITLES PUBLISHED BY ARETEEM PRESS

**Cracking the High School Math Competitions (and Solutions Manual)** - Covering AMC 10 & 12, ARML, and ZIML
**Mathematical Wisdom in Everyday Life (and Solutions Manual)** - From Common Core to Math Competitions
**Geometry Problem Solving for Middle School (and Solutions Manual)** - From Common Core to Math Competitions
**Fun Math Problem Solving For Elementary School (and Solutions Manual)**

ZIML MATH COMPETITION BOOK SERIES

**ZIML Math Competition Book Division E 2016-2017**
**ZIML Math Competition Book Division M 2016-2017**
**ZIML Math Competition Book Division H 2016-2017**
**ZIML Math Competition Book Jr Varsity 2016-2017**
**ZIML Math Competition Book Varsity Division 2016-2017**
**ZIML Math Competition Book Division E 2017-2018**
**ZIML Math Competition Book Division M 2017-2018**
**ZIML Math Competition Book Division H 2017-2018**
**ZIML Math Competition Book Jr Varsity 2017-2018**
**ZIML Math Competition Book Varsity Division 2017-2018**

MATH CHALLENGE CURRICULUM TEXTBOOKS SERIES

**Math Challenge I-A Pre-Algebra and Word Problems**
**Math Challenge I-B Pre-Algebra and Word Problems**
**Math Challenge I-C Algebra**
**Math Challenge II-A Algebra**
**Math Challenge II-B Algebra**
**Math Challenge III Algebra**
**Math Challenge I-A Geometry**
**Math Challenge I-B Geometry**
**Math Challenge I-C Topics in Algebra**
**Math Challenge II-A Geometry**
**Math Challenge II-B Geometry**

**Math Challenge III Geometry**
**Math Challenge I-B Counting and Probability**
**Math Challenge II-A Combinatorics**
**Math Challenge I-B Number Theory**
**Math Challenge II-A Number Theory**

COMING SOON FROM ARETEEM PRESS

**Fun Math Problem Solving For Elementary School Vol. 2 (and Solutions Manual)**
**Counting & Probability for Middle School (and Solutions Manual)** - From Common Core to Math Competitions
**Number Theory Problem Solving for Middle School (and Solutions Manual)** - From Common Core to Math Competitions
Other volumes in the **Math Challenge Curriculum Textbooks Series**

The books are available in paperback and eBook formats (including Kindle and other formats).
To order the books, visit https://areteem.org/bookstore.

# Contents

# Introduction

Each month during the school year, Areteem Institute hosts the online Zoom International Math League (ZIML) competitions. Students can compete in one of five divisions based on their age and mathematical level (details shown on Page 9).

This book contains the problems, answers, and full solutions from the nine ZIML Division H Competitions held during the 2017-2018 School Year. It is divided into three parts:

1. The complete Division H ZIML Competitions (20 questions per competition) from October 2017 to June 2018.
2. The solutions for each of the competitions, including detailed work and helpful tricks.
3. An appendix including the topics and knowledge points covered for Division H, a glossary including common mathematical terms, and answer keys for each of the competitions so students can easily check their work.

The questions found on the ZIML competitions are meant to test your problem solving skills and train you to apply the knowledge you know to many different applications. We hope you enjoy the problems!

## About Zoom International Math League

The Zoom International Math League (ZIML) has a simple goal: provide a platform for students to build and share their passion for math and other STEM fields with students from around the globe. Started in 2008 as the Southern California Mathematical Olympiad, ZIML has a rich history of past participants who have advanced to top tier colleges and prestigious math competitions, including American Math Competitions, MATHCOUNTS, and the International Math Olympaid.

The ZIML Core Online Programs, most available with a free account at ziml.areteem.org, include:

- **Daily Magic Spells:** Provides a problem a day (Monday through Friday) for students to practice, with full solutions available the next day.
- **Weekly Brain Potions:** Provides one problem per week posted in the online discussion forum at ziml.areteem.org. Usually the problem does not have a simple answer, and students can join the discussion to share their thoughts regarding the scenarios described in the problem, explore the math concepts behind the problem, give solutions, and also ask further questions.
- **Monthly Contests:** The ZIML Monthly Contests are held the first weekend of each month during the school year (October through June). Students can compete in one of 5 divisions to test their knowledge and determine their strengths and weaknesses, with winners announced after the competition.
- **Math Competition Practice:** The Practice page contains sample ZIML contests and an archive of AMC-series tests for online practice. The practices simulate the real contest environment with time-limits of the contests automatically controlled by the server.
- **Online Discussion Forum:** The Online Discussion Forum

is open for any comments and questions. Other discussions, such as hard Daily Magic Spells or the Weekly Brain Potions are also posted here.

These programs encourage students to participate consistently, so they can track their progress and improvement each year.

In addition to the online programs, ZIML also hosts onsite Local Tournaments and Workshops in various locations in the United States. Each summer, there are onsite ZIML Competitions at held at Areteem Summer Programs, including the National ZIML Convention, which is a two day convention with one day of workshops and one day of competition.

ZIML Monthly Contests are organized into five divisions ranging from upper elementary school to advanced material based on high school math.

- **Varsity:** This is the top division. It covers material on the level of the last 10 questions on the AMC 12 and AIME level. This division is open to all age levels.
- **Junior Varsity:** This is the second highest competition division. It covers material at the AMC 10/12 level and State and National MathCounts level. This division is open to all age levels.
- **Division H:** This division focuses on material from a standard high school curriculum. It covers topics up to and including pre-calculus. This division will serve as excellent practice for students preparing for the math portions of the SAT or ACT. This division is open to all age levels.
- **Division M:** This division focuses on problem solving using math concepts from a standard middle school math curriculum. It covers material at the level of AMC 8 and School or Chapter MathCounts. This division is open to all students who have not started grade 9.

- **Division E:** This division focuses on advanced problem solving with mathematical concepts from upper elementary school. It covers material at a level comparable to MOEMS Division E. This division is open to all students who have not started grade 6.

This problem book features the Division H Contests. For a detailed list of topics covered for Division H see p.159 in the Appendix.

## About Areteem Institute

Areteem Institute is an educational institution that develops and provides in-depth and advanced math and science programs for K-12 (Elementary School, Middle School, and High School) students and teachers. Areteem programs are accredited supplementary programs by the Western Association of Schools and Colleges (WASC). Students may attend the Areteem Institute through these options:

- Live and real-time face-to-face online classes with audio, video, interactive online whiteboard, and text chatting capabilities;
- Self-paced classes by watching the recordings of the live classes;
- Short video courses for trending math, science, technology, engineering, English, and social studies topics;
- Summer Intensive Camps on prestigious university campuses and Winter Boot Camps;
- Practice with selected daily problems for free, and monthly ZIML competitions at ziml.areteem.org.

The Areteem courses are designed and developed by educational experts and industry professionals to bring real world applications into STEM education. The programs are ideal for students who wish to build their mathematical strength in order to excel academically and eventually win in Math Competitions (AMC, AIME, USAMO, IMO, ARML, MathCounts, Math Olympiad, ZIML, and other math leagues and tournaments, etc.), Science Fairs (County Science Fairs, State Science Fairs, national programs like Intel Science and Engineering Fair, etc.) and Science Olympiad, or purely want to enrich their academic lives by taking more challenges and developing outstanding analytical, logical thinking and creative problem solving skills.

Since 2004 Areteem Institute has been teaching with methodology that is highly promoted by the new Common Core State Standards: stressing the conceptual level understanding of the math concepts, problem solving techniques, and solving problems with real world applications. With the guidance from experienced and passionate professors, students are motivated to explore concepts deeper by identifying an interesting problem, researching it, analyzing it, and using a critical thinking approach to come up with multiple solutions.

Thousands of math students who have been trained at Areteem achieved top honors and earned top awards in major national and international math competitions, including Gold Medalists in the International Math Olympiad (IMO), top winners and qualifiers at the USA Math Olympiad (USAMO/JMO), and AIME, top winners at the Zoom International Math League (ZIML), and top winners at the MathCounts National. Many Areteem Alumni have graduated from high school and gone on to enter their dream colleges such as MIT, Cal Tech, Harvard, Stanford, Yale, Princeton, U Penn, Harvey Mudd College, UC Berkeley, UCLA, etc. Those who have graduated from colleges are now playing important roles in their fields of endeavor.

Further information about Areteem Institute, as well as updates and errata of this book, can be found online at http://www.areteem.org.

## Acknowledgments

This book contains the Online ZIML Division H Problems from the 2017-18 school year. These problems were created and compiled by the staff of Areteem Institute. These problems were inspired by questions from the Areteem Math Challenge Courses, past questions on the ACT/SAT/GRE, past math competitions, math textbooks, and countless other resources and people encountered by the Areteem Curriculum Department in their life devoted to math. We thank all these sources for growing and nurturing our passion for math.

The Areteem staff, including John Lensmire, David Reynoso, Kevin Wang, and Kelly Ren, are the main contributors who compiled, edited, and reviewed this book.

Lastly, thanks to all the students who have participated and continue to participate in the Zoom International Math League. Your dedication to the Daily Magic Spells and Monthly Contests makes all of this possible, and we hope you continue to enjoy ZIML for years to come!

# 1. ZIML Contests

This part of the book contains the Division H ZIML Contests from the 2017-18 School Year. There were nine monthly competitions, held on the dates found below:

- October 6-8
- November 3-5
- December 1-3
- January 5-7
- February 2-4
- March 2-4
- April 6-8
- May 4-6
- June 1-3

## 1.1  ZIML October 2017 Division H

Below are the 20 Problems from the Division H ZIML Competition held in October 2017.
The answer key is available on p.172 in the Appendix.
Full solutions to these questions are available starting on p.78.

### Problem 1
There are 100 total students in William's grade at school. At the school award ceremony, awards were given out to (i) everyone who played a sport, and (ii) everyone on the honor roll. Every student got at least one award! If 80 of Williams classmates played a sport and 60 were on the honor roll, what is the chance that someone who played a sport was also on the honor roll? Express your answer as $K\%$, with $K$ rounded to the nearest integer.

### Problem 2
Solve the system of equations

$$\begin{cases} \dfrac{1}{2}\sqrt{x} - 2y &= 4, \\ \dfrac{1}{2}\sqrt{x} + 2y &= 6. \end{cases}$$

There is one solution of the form $(A, B)$. What is $A + B$, rounded to the nearest tenth if necessary?

## Problem 3

Consider the parallelogram $ABCD$. Let $E$ and $F$ be midpoints of the sides $CD$ and $DA$, respectively.

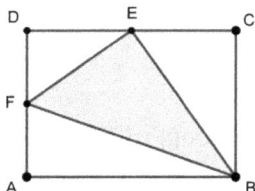

If the area of $ABCD$ is 20, what is the area of $BEF$? Round your answer to the nearest tenth if necessary.

## Problem 4

Mark $P$ inside square $ABCD$, so that triangle $ABP$ is equilateral. Find the size of $\angle CDP$ in degrees.

## Problem 5

Troy shoots an arrow while his friends record it from afar. After looking at the recording, they noticed that the arrow followed the path of a parabola. If it took 2.3 seconds for the arrow to reach its highest point, how many seconds passed after shooting the arrow until it reached the same height it was launched from?

## Problem 6

Mark wrote 6 different numbers, one each on the front and back of 3 cards, and laid the cards on a table, as shown.

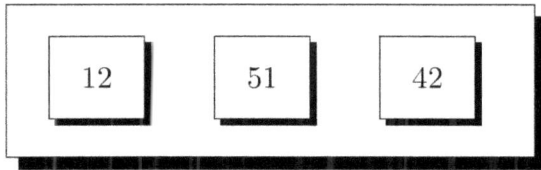

The three numbers on the back of the cards are prime numbers. Adding the front and back numbers gives the same result for all three cards. What is the average of the hidden prime numbers?

## Problem 7

How many of the first 200 positive integers $(1, 2, \ldots, 200)$ are divisible by 4 or 6?

## Problem 8

A cubic equation $y = ax^3 + bx^2 + cx + d$ has a root of $x = -1$ and a double root of $x = 2$. If the $y$-intercept is $y = 8$, what is $a + b + c + d$?

## Problem 9

Suppose $A, B, C$ are points on a circle such that the angular measures of arc $AB$ (not containing $C$) and arc $BC$ (not containing $A$) are in ratio $2 : 3$. Suppose further that $\angle ABC = 80°$. Find the measure of $\angle BAC$.

### Problem 10

What is the smallest integer value $K$ so that the line $y = x + K$ intersects the parabola $y = x^2 - 5x + 14$ at least once?

### Problem 11

Find the smallest positive angle $\theta$ (measured in degrees) such that $2\sin(2\theta - 90°) = \sqrt{2}$. Round your answer to the nearest hundredth if necessary.

### Problem 12

$N$ is a positive integer such that $N$ leaves a remainder of 3 when divided by 5, leaves a remainder of 5 when divided by 7, and leaves a remainder of 7 when divided by 9. What is the smallest such $N$?

### Problem 13

Suppose you have circle $(x - 2)^2 + (y + 2)^2 = 25$ with center $C$ and line $x = 4$. This line intersects the circle at 2 points, call them $A$ and $B$. Use points $A$, $B$, and $C$ to form a triangle $\triangle ABC$. The area of this triangle can be written as $R\sqrt{S}$ in simplest radical form. What is $R + S$?

### Problem 14

$p + i\sqrt{2}$ and $2 - i\sqrt{q}$ (for integers $p$ and $q$) are solutions to the quadratic equation $3x^2 - 12x + C = 0$ where $C$ is an integer. What is $C$?

## Problem 15

The equation $\log_3(\log_2|x^2 - 9|) = 0$ has 4 solutions that can be written as $\pm\sqrt{K}$ and $\pm\sqrt{L}$ for integers $K \neq L$. What is $K + L$?

## Problem 16

The equation $6^x + 8^x = 5392$ has one integer solution $x = L$. What is $L$?

## Problem 17

Points $A$, $B$, $C$ are on a straight line with $AB = 6$ and $BC = 2$. Point $D$ is such that $AD = 8$ and $BD = 4$. What is $CD$? Round your answer to the nearest integer if necessary.

## Problem 18

Consider an isosceles triangle cut out of a semicircle as in the diagram below.

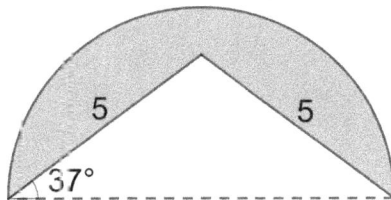

If we use the approximations $\sin(37°) = 0.6$ and $\pi = 3.1$, what is the area of the shaded region?

## Problem 19

$x$ is a real number satisfying $x^2 - 3x - 5 = 0$. Calculate $x^4 - 2x^3 - 9x^2 - 2x + 8$. Round your answer to the nearest hundredth if necessary.

## Problem 20

Two leadership positions for the school math club are being chosen from a group of three girls and five boys. The probability that at one girl and one boy are chosen can be written as $\dfrac{P}{Q}$ for positive $P, Q$ with $\gcd(P, Q) = 1$. What is $P + Q$?

## 1.2   ZIML November 2017 Division H

Below are the 20 Problems from the Division H ZIML Competition held in November 2017.

The answer key is available on p.173 in the Appendix.

Full solutions to these questions are available starting on p.86.

### Problem 1

$ABCD$ is a trapezoid and $\overline{AD}$ and $\overline{BC}$ are extended to intersect at $E$ as shown below.

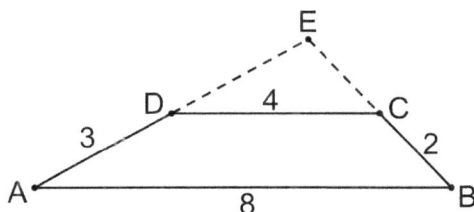

What is the perimeter of triangle $CDE$?

### Problem 2

Consider solutions to $x^4 - 7x^3 + x^2 + 15x + 6 = 0$. Two solutions can be written in the form $A \pm \sqrt{B}$ for integers $A, B$, and $B$ NOT a perfect square. What is $A + B$?

### Problem 3

How many distinct rearrangements of the letters in BANANAS are there?

## Problem 4

Find the global maximum of $f(x) = 61 + 72x - 36x^2$. Round your answer to the nearest integer if necessary.

## Problem 5

A dog is leashed at the top corner of a building whose base is an equilateral triangle with side length 3 m, as shown in the diagram.

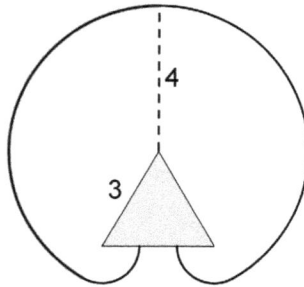

The length of the rope is 4 m. The total area of the region (in m$^2$) that the dog can reach can be expressed as $K \times \pi$ for a positive $K$. What is $K$, rounded to the nearest integer if necessary?

## Problem 6

What is the tens digit of $6^{2018}$?

## Problem 7

Consider the sector shown below with a radius of 6.

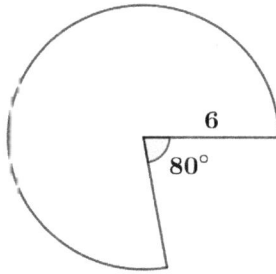

For this problem assume $\pi = 3$ and calculate the perimeter of the sector. Round your answer to the nearest integer if necessary.

## Problem 8

Find $m$ such that the equation $x^2 + 3x - m = -6$ has exactly one solution. Round your answer answer to the nearest hundredth if necessary.

## Problem 9

Larry and Moe love betting. Larry bets Moe that if they keep flipping a fair coin (with outcomes heads or tails) that there will be 6 heads before there are 6 tails. Currently there are 5 heads and 3 tails. The probability that Larry wins his bet can be written as $K\%$. What is $K$, rounded to the nearest integer if necessary?

## Problem 10

Consider the domain of $y = \sqrt{17x - x^2}$. How many integers are there in the domain?

## Problem 11

Everyday at work Peter drinks from a cylindrical cup with a height of 8 cm and whose base has a radius of 3 cm. He has a mixing stick of length 15 cm which he rests in the glass as shown in the diagram below:

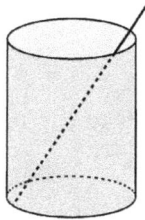

How far in cm does the mixing stick extend out of the cup? Round your answer to the nearest cm if necessary.

## Problem 12

Recall $\lfloor x \rfloor$ is the greatest integer $\leq x$ and $\{x\} = x - \lfloor x \rfloor$.

The equation $2 \lfloor x \rfloor - x = \{x\}$ has one solution where $x \neq 0$. What is this solution? Round your answer to the nearest hundredth if necessary.

## Problem 13
The equation $\log_3(x) - \log_9(x) + \log_{27}(x) = \dfrac{22}{3}$ has one integer solution. Find this solution.

## Problem 14
$\triangle ABC$ has sides $BC = 5$ with $\angle A = 30°, \angle B = 45°$. What is side $AC$? Round your answer to the nearest tenth if necessary.

## Problem 15
Suppose you have a list of 9 elements where each element is an non-negative integer less than or equal to 20. What is the largest possible difference between the median and mean of this list? Round your answer to the nearest tenth if necessary.

## Problem 16
There is one solution to $3\cos(\theta) = 2\sin^2(\theta)$ for $0° \leq \theta \leq 90°$. What is this solution in degrees? Round your answer to the nearest degree if necessary.

## Problem 17
Let $ABCD$ be a parallelogram, and $E, H, F, G$ be points on sides $\overline{AB}, \overline{BC}, \overline{CD}, \overline{DA}$ respectively, and $\overline{EF} \| \overline{BC}$ and $\overline{GH} \| \overline{AB}$. Let $P$ be the intersection of $\overline{EF}$ and $\overline{GH}$.

If the areas of $GPFD, PHCF, EBHP$ are respectively $12, 8, 20$ (written $[GPFD] = 12, [PHCF] = 8, [EBHP] = 20$), find the area of $ABCD$.

## Problem 18

The product of all the factors of $2^{11}$ is $2^M$ for $M$ an integer. What is $M$?

## Problem 19

If you calculate $(1 - i\sqrt{3})^{10}$ you will get $K \times (1 - i\sqrt{3})$ for an integer $K$. What is $K$?

## Problem 20

Let $ABCD$ be a rectangle with $BC = 2$ and $CD = 1$. Construct equilateral triangles $BCE$ and $CDF$ on the sides of the rectangle. What is $EF^2$ rounded to the nearest integer?

## 1.3   ZIML December 2017 Division H

Below are the 20 Problems from the Division H ZIML Competition held in December 2017.

The answer key is available on p.174 in the Appendix.

Full solutions to these questions are available starting on p.95.

### Problem 1
Solve

$$\frac{x+3}{x-3} - \frac{x+2}{x-2} = 1.$$

What is the sum of all the real solutions? Round your answer to the nearest integer if necessary.

### Problem 2
Adam and Alice, Bob and Bridget, Charlie and Claire, and Dan and Daisy are 4 couples who plan to go to a movie together. They reserved 8 seats at the movie theater all in a row. If everyone sits next to their partner during the movie, how many different seating arrangements are there?

### Problem 3
Suppose $O$ is the center of a circle with radius 4. Let $ABCO$ be a rhombus with $A, B, C$ on the circle. What is the area of the rhombus rounded to the nearest integer?

### Problem 4
Let $f(x) = x^4 - 13x^2 + 12$. For how many integers $n$ is $f(n) \leq 0$?

## Problem 5

Consider the line $y = 3x - 10$. The closest this line gets to the origin is $\sqrt{D}$. What is $D$?

## Problem 6

How many ways can you place four identical checkers on the outer edges of an $8 \times 8$ checkerboard so that the checkers are in different rows and in different columns?

## Problem 7

How many zeros are at the end of 200!?

Recall $200! = 200 \times 199 \times 198 \times \cdots \times 2 \times 1$.

## Problem 8

A circle with center $(3,4)$ contains the point $(5,-2)$. The equation of this circle can be expressed in the form

$$x^2 + y^2 = Ax + By + C,$$

where $A$, $B$ and $C$ are integers. What is $C$?

## Problem 9

The equation $\log_3(x) + 2\log_9(x) + 3\log_{27} = 9$ has one real solution for $x$. What is this solution? Round your answer to the nearest tenth if necessary.

## Problem 10

Let points $O = (0,0)$ and $Q = (2\sqrt{2},0)$. Point $P$ lies on the parabola $y = 4 - x^2$ so that $OP = PQ$. The area of triangle $OPQ$ is $\sqrt{K}$ for an integer $K$, what is $K$?

## Problem 11

Peter and Paul each roll a fair six-sided die. The probability that Peter's roll is greater than or equal to Paul's roll can be written as $\dfrac{P}{Q}$ for positive integers $P, Q$ with $\gcd(P,Q) = 1$. What is $P+Q$?

## Problem 12

One of the real solutions of $|x - |2x + 1|| = 3$ is of the form $\dfrac{P}{Q}$, where $Q > 1$ and $\gcd(P,Q) = 1$. What is $Q - P$?

## Problem 13

Suppose in $\triangle ABC$ (with sides $a, b, c$ opposite from $\angle A, \angle B, \angle C$) we know $a = 5$, $\angle A = 30°$, and $\angle B = 45°$. What is side $b$? If necessary round your answer to the nearest tenth.

## Problem 14

If $0° < x < 90°$ and $4\sin^2(x) = 3 - \sin(x)$, what is the value of $\cos^2(x)$? Express your answer as a decimal rounded to the nearest hundredth if needed.

## Problem 15
How many positive integers $\leq 500$ have exactly 3 factors?

## Problem 16
Let $f(x) = 20 - cx^3$, where $c$ is a constant. If $f(3) = 15$, what is the value of $f(-3)$?

## Problem 17
Let $ABCD$ be a trapezoid with bases $\overline{AB} \parallel \overline{CD}$. If $AB = 31$, $AD = 15$, $CD = 6$, and $\sin A = \frac{4}{5}$, what is the perimeter of $ABCD$?

## Problem 18
Let $f(z) = 6z^5 + 5z^4 + 4z^3 + 3z^2 + 2z + 1$. Then $f(i) = A + Bi$ for integers $A, B$. What is $A^2 + B^2$? Here $i = \sqrt{-1}$.

## Problem 19
The line $y = mx$ (with $m > 0$) forms a 20° angle with the $x$-axis. The line $y = x - 2$ forms a 45° angle with the $y$-axis. The lines $y = mx$ and $y = x - 2$ meet and form an acute angle of $K°$. What is $K$?

## Problem 20

Suppose an uncle distributes $1 bills among his nieces and nephews. If he distributed the dollars evenly among the nieces, each niece would get 44 dollars, and if he distributed them evenly among the nephews, each nephew would get 77 dollars. In fact, he distributes the dollars evenly among all his nieces and nephews. How many dollars does each niece or nephew receive?

## 1.4   ZIML January 2018 Division H

Below are the 20 Problems from the Division H ZIML Competition held in January 2018.
The answer key is available on p.175 in the Appendix.
Full solutions to these questions are available starting on p.104.

### Problem 1
Sara bought a new car. worth $15,000. After some research, she determined that for each year she owned the car, the car would be worth 20% less than the year before. How much is Sara's car worth after 3 years? Give your answer in dollars, rounded to the nearest whole number.

### Problem 2
Consider the equation $3x^2 + bx + 12 = 0$. For how many integers $b$ does the equation have no real solutions?

### Problem 3
Find the imaginary part of $\dfrac{4+i}{3-i}$. Express your answer as a decimal, rounded to the nearest hundredth if necessary.

### Problem 4
When the numbers 513, 571, and 658 are divided by an integer $D > 1$, they all have the same remainder $R$. What is this remainder?

---

## Problem 5

$\triangle ABC \sim \triangle DEF$. If $AB = 20$, $BC = 12$, $DE = x+2$, and $EF = x-2$, what is $x$?

## Problem 6

For his New Year's Party, Peter had a message written out with one letter on each piece of paper as shown below

$$\boxed{H}\,\boxed{A}\,\boxed{P}\,\boxed{P}\,\boxed{Y}\,\boxed{N}\,\boxed{E}\,\boxed{W}\,\boxed{Y}\,\boxed{E}\,\boxed{A}\,\boxed{R}$$

After the party, Peter randomly grabbed 2 sheets of paper. How many different collections of sheets could he have? The order he grabs the sheets is not important, so the collection $\boxed{H}\,\boxed{A}$ is the same as $\boxed{A}\,\boxed{H}$. Assume all the repeated letters are identical.

## Problem 7

What is the units digit of $19^{19} + 99^{99}$?

## Problem 8

Suppose you have a trapezoid $ABCD$ with $\overline{AB}$ parallel to $\overline{CD}$. Let $E$ be the intersection of the diagonals. Suppose $AB = 3, CD = 5$ and $[ADE] = 15$ (denoting $\triangle ADE$ has area 15). Find the area of $ABCD$.

## Problem 9

In the $6 \times 6 \times 6$ cubic figure below, if the path is required to be along the surface of the cube, the length of the shortest path from point $A$ to $B$ can be expressed in the form $\sqrt{K}$ for an integer $K$. What is $K$?

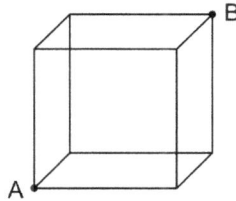

## Problem 10

The equation $x^3 - 5x^2 + 5x + 3 = 0$ has one solution of $x = 3$ and the others in the form $A \pm \sqrt{B}$ where $A$ and $B$ are integers. What is $A + B$?

## Problem 11

How many factors of 4000 are perfect squares? Remember that 1 and 4000 are both factors of 4000.

## Problem 12

Suppose you start at a flag and walk 15 feet north. You then walk west a while before stopping so that you are $30°$ west of the flag. How far are you from the flag? Round your answer to the nearest integer. (Note $30°$ west of the flag means the line connecting you to the flag forms a $30°$ angle with a line straight north.)

## Problem 13

Laura use machine learning to write a program to help classify squares. When shown a square, Laura's program correctly identifies the shape as a square 95% of the time. When shown a non-square shape, Laura's program correctly identifies it as not a square 90% of the time. Laura has a data set of 1000 shapes, 200 of which are squares (the rest are not squares). If Laura randomly chooses one shape from her data set and runs her program on it, then the chance the program identifies it (correctly or incorrectly) as a square can be written as $P\%$. What is $P$, rounded to the nearest integer if necessary?

## Problem 14

Simplify the sum

$$\frac{1}{\sqrt{1}+\sqrt{2}}+\frac{1}{\sqrt{2}+\sqrt{3}}+\frac{1}{\sqrt{3}+\sqrt{4}}+\cdots+\frac{1}{\sqrt{99}+\sqrt{100}}.$$

Round your answer to the nearest tenth if necessary.

## Problem 15

Consider pairs $(x,y)$ satisfying (i) $2|x|+y=5$ and (ii) $|x|-y=1$. There is one pair such that $x \cdot y < 0$. What is $x \cdot y$ for this pair?

## Problem 16

Suppose $\widehat{AB}$ and $\widehat{CD}$ are arcs each with angular size $50°$ and if rays $\overrightarrow{BA}, \overrightarrow{DC}$ are extended to intersect at a point $E$ (so $A$ is on $\overline{BE}$ and on $\overline{DE}$), $\angle AEC = 50°$. Find the angular size of arc $\widehat{BD}$. Express your answer in degrees, rounded to the nearest tenth if necessary.

## Problem 17

Consider a circle divided into 5 regions as in the diagram below, where the size of each region is proportional to its number.

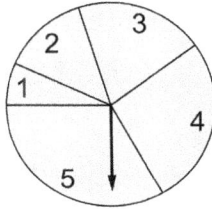

The spinner is spun around the center of the circle and randomly stops in one of the regions. The probability the spinner stops in the region labeled with an odd number is $\dfrac{N}{M}$ where the fraction is in lowest terms. What is $M - N$?

## Problem 18

Four non-overlapping regular plane polygons all have sides of length 3. The polygons meet at a point $A$ in such a way that the sum of the four interior angles at $A$ is $360°$. Among the four polygons, two are squares and one is a triangle, as shown below, except for the missing fourth polygon.

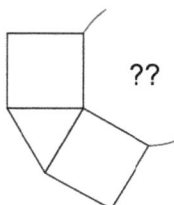

With the fourth polygon included, what is the perimeter of the entire shape?

## Problem 19

Suppose John's cup is a cylinder with diameter 5 cm and height 12 cm. Suppose he wants to fill his cup halfway using a water fountain, tipping it over slightly to allow water to enter the cup. The water comes out of the fountain at a height $h$. What is the smallest integer height, $h$, (in cm) that allows John to fill his cup halfway as wanted?

## Problem 20

What is the sum of all the solutions to $(\log_4 x)^2 + 2 = \log_4(x^3)$? Round your answer to the nearest integer if necessary.

## 1.5   ZIML February 2018 Division H

Below are the 20 Problems from the Division H ZIML Competition held in February 2018.
The answer key is available on p.176 in the Appendix.
Full solutions to these questions are available starting on p.112.

### Problem 1
If $\log_2(x) + \log_2(x^2) = 15$ for an integer $x$, what is $x$?

### Problem 2
Suppose you have 6 men and 6 women at a dance class. How many ways are there to divide the class into pairs (the pairs have no particular order to them) if each pair consists of one man and one woman?

### Problem 3
A rectangular prism has one face with area 35 and another face with area 42. If all dimensions of the prism are of integer lengths greater than 1, what is the surface area of the rectangular prism?

### Problem 4
Points $A, B, C$ are all on the same line with $A = (1,3)$, $B = (3,k)$, and $C = (k,7)$ for some number $k > 0$. What is $k$?

## Problem 5

$\theta$ is an acute angle in a right triangle and $\sin(\theta) = \dfrac{12}{19}$. Then $\cos(\theta)$ can be approximated by $\dfrac{P}{19}$ where $P$ is rounded to the nearest integer. What is $P$?

## Problem 6

Find the smallest 3-digit number that leaves a remainder of 1 when divided by 7 and 5 when divided by 16.

## Problem 7

You have a list of 5 numbers, chosen from the integers $1, 2, 3, \ldots, 20$ with no repeats. How many different averages of the 5 numbers are possible?

## Problem 8

There are two integer roots and two irrational roots to the equation $x^4 - x^3 - 5x^2 + 3x + 2 = 0$. The irrational roots can be written the form $A \pm \sqrt{B}$ for integers $A, B$. What is $A + B$?

## Problem 9

A triangle has a $45°$ angle and a $30°$ angle, and the side opposite the $45°$ angle has length 12. The side opposite the $30°$ angle has length $\sqrt{K}$ for an integer $K$. What is $K$?

## Problem 10

You are standing 5 ft away from a wall where a huge painting is being displayed. If you look down at an angle of $40°$ you are staring right at the bottom of the painting. If you look up at an angle of $65°$ you are staring right at the top of the painting. If $\tan(40°) \approx 0.84$, and $\tan(65°) \approx 2.14$, what is the height of the painting?

## Problem 11

For a complex number $z$, $\bar{z}$ denotes the complex conjugate of $z$. Consider all the complex numbers of the form $z = a + bi$ with $a, b$ integers such that $1 \le a^2 + b^2 \le 100$. How many such numbers satisfy $z^2 = \bar{z}^2$?

## Problem 12

The slots of a roulette wheel are numbered 1-36, 0 and 00. The numbers 0 and 00 are green. For numbers in the ranges 1-10 and 19-28, the odd numbers are red and even numbers are black. For numbers in the ranges 11-18 and 29-36, the odd numbers are black and even numbers are red. The probability a randomly chosen slot is both odd and black can be written as $\dfrac{P}{Q}$ for positive integers $P, Q$ with $\gcd(P, Q) = 1$. What is $Q - P$?

## Problem 13

$ABCD$ is a square with side length 10. Draw a semicircle with diameter $AB$ inside the square. Let $F$ be a point of $\overline{AD}$ so that $\overline{CF}$ is tangent to the semicircle (at point $E$). What is the length of $CF$? Round your answer to the nearest tenth if necessary.

## Problem 14

Let $f(x) = \frac{1}{3}x - 3$. Find the sum of all possible values of $x$ satisfying $f(x) \cdot f^{-1}(x) = 0$. Round your answer to the nearest integer if necessary. (Here $f^{-1}(x)$ denotes the inverse of $f(x)$.)

## Problem 15

Find all ordered pairs $(x, y)$ of prime numbers satisfying the equation $x(x+y) = 120$. What is the sum of all possible $y$? (If there are no such $y$, input 0 for your answer.)

## Problem 16

Let $x, y = 2 \pm \sqrt{7}$. The expression $\dfrac{x^2 + 2xy + y^2}{x^3y^2 + x^2y^3 + x + y}$ can be written as $\dfrac{P}{Q}$ for integers $P$ and $Q$ with $Q > 0$ and $\gcd(P, Q) = 1$. What is $P + Q$?

## Problem 17

Consider points $A = (1, 3)$ and $B = (4, 2)$. If $\triangle ABC$ is isosceles with $AC = BC$, then $C$ must lie on the line $y = mx + b$ for real numbers $m$ and $b$. What is $m \times b$? Round your answer to the nearest integer if necessary.

## Problem 18

The product of all the factors of $2^{15}$ is $2^M$ for $M$ in integer. What is $M$?

## Problem 19

Two tangent circles are drawn with centers $A$ and $B$ in isosceles right triangle $\triangle ABC$.

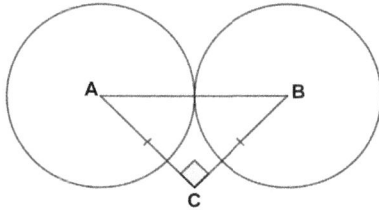

Using the approximation $\pi \approx \dfrac{22}{7}$ the area inside the triangle but outside the two circles is $\dfrac{P}{Q}$ of the entire triangle. What is $P + Q$?

## Problem 20

The equation $\dfrac{1}{\sqrt{x+2}} + \sqrt{x+2} = \dfrac{10}{3}$ has one solution of the form $\dfrac{P}{Q}$ for $P$ an integer and $Q > 1$ with $\gcd(P, Q) = 1$. What is $P - Q$?

## 1.6   ZIML March 2018 Division H

Below are the 20 Problems from the Division H ZIML Competition held in March 2018.
The answer key is available on p.177 in the Appendix.
Full solutions to these questions are available starting on p.121.

### Problem 1
The line $x - y = 10$ intersects the parabola $x^2 - 3y = 28$ at two points. What is the sum of the $y$-coordinates of these points?

### Problem 2
The four lines $y = x$, $y = x + 2$, $y = 2x$, and $y = 2x + 2$ form the boundary of a quadrilateral. What is the area of this quadrilateral? Round your answer to the nearest tenth if necessary.

### Problem 3
Expand $(5x - 3)^8$ to get $ax^8 + bx^7 + \cdots + +hx + i$ for integers $a, b, \ldots, i$. What is the sum $a + b + \cdots + i$?

## Problem 4

Consider the trapezoid with vertices $(0,0)$, $(4,0)$, $(4,2)$, and $(0,5)$ shown below.

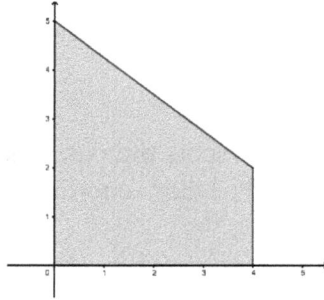

This trapezoid is rotated about the $y$-axis to form a three-dimensional solid. The volume of this solid can be expressed as $L \times \pi$. What is $L$, rounded to the nearest integer if necessary?

## Problem 5

Last month Bruce helped 41 customers and sold a total of $662 worth of merchandise. Assume the amount sold to each customer was $\geq \$10$ and an integer. Bruce gets a bonus based on the median value of his sales. What is the difference between the largest possible bonus and smallest possible bonus Bruce could receive for last month's sales? Give your answer in dollars, rounded to the nearest hundredth if necessary.

## Problem 6

How many numbers from $1, 2, 3, 4, \ldots, 300$ have exactly 3 factors?

## Problem 7

Square $DEFG$ is drawn inside the 30-60-90 triangle $\triangle ABC$ as shown below.

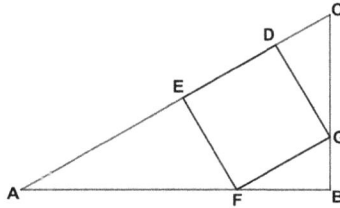

If the area of square $DEFG$ is 36, then length $AB$ can be written as $P+Q\sqrt{R}$ for integers $P,Q,R$ so that $R$ contains no squares as factors. What is $P+Q+R$?

## Problem 8

Peter graphs a degree 4 polynomial $f(x)$. He forgot the exact equation for $f(x)$, but he remembered that $f(5) = 18$ and that the equation did not contain a cubic $(x^3)$ or linear $(x)$ term. What is $f(-5)$?

## Problem 9

$\triangle ABC$ has 3 sides that are all integers. They are of length

$$AB = 20, AC = 35, \text{ and } BC = \frac{z}{3}.$$

How many values of $z$ are possible?

## Problem 10

Find the sum of all possible values of $x$ such that

$$\log_x(x-2) + \log_x(3x+5) = 2.$$

Round your answer to the nearest tenth if necessary. (Input an answer of 0 if there are no solutions.)

## Problem 11

A point is randomly picked inside the large triangle below that is divided into 10 regions. (Each of the triangles is equilateral and they are divided using the midpoints of the sides.)

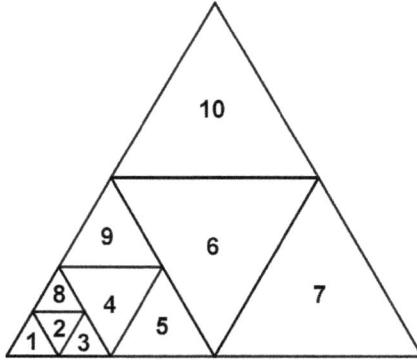

The probability the point belongs to a triangle labeled with an even number can be written as $\dfrac{R}{S}$ for positive integers $R, S$ with $\gcd(R, S) = 1$. What is $R + S$?

## Problem 12

A dog is tied to the ground with a leash of length 10 feet. He tries to run around a nearby tree with diameter 3 feet, but gets stuck along the way (as the leash is too short).

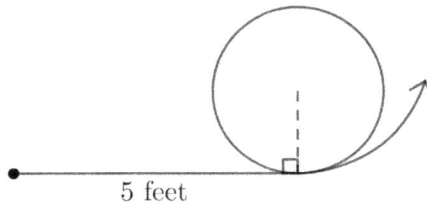

5 feet

Using the approximation $\pi \approx 3$, the dog is able to run around $K\%$ of the tree. What is $K$ rounded to the nearest integer?

## Problem 13

Two digits $A, B$ are chosen and used to form the 8-digit number $\overline{20AB18BA}$. What is the largest possible such number that is divisible by 18?

## Problem 14

$D$ is the smallest positive integer such that $x = D°$ is a solution to $6\sin^2(2x) = 2 - 2\sin(x)\cos(x)$. What is $D$?

## Problem 15

Consider the equation $\dfrac{x}{5} + \dfrac{y}{7} = \dfrac{57}{35}$, with $x$ and $y$ integers. What is the largest possible value of $x^2 - y^2$ if $(x, y)$ is a solution to the equation?

## Problem 16

Points $A$, $B$, $C$, and $D$ are on a circle and form quadrilateral $ABCD$. If $\angle A = 2x$, $\angle B = 3x$, $\angle C = 5y$, and $\angle D = 3y$, what is $x$ in degrees? Round your answer to the nearest integer if necessary.

## Problem 17

It is time for Dennis to make a new password. He's not too creative so he decides to create a password with his name and birthday, which is May 25. He wants to use the letters of his name (in order) and the letters/symbols of his birthday (in order). That is, the password could start with either D or M; if it starts with D the next letter could be M or e, and if it starts with M the next letter could be D or a. Examples of possible passwords are MayDennis25 or DenMaynis25 or DeManyni2s5. A password of 25DennisMay or nisMay25Den is NOT allowed. How many possible passwords could Dennis make?

## Problem 18

What is the sum of all solutions to $\sqrt{x+3} - \sqrt{3x-2} = -1$? Round your answer to the nearest tenth if necessary. (Input an answer of 0 if there are no solutions.)

## Problem 19

In acute $\triangle ABC$, suppose $AC = 24$, $BC = 26$ and $\sin C = \dfrac{4\sqrt{3}}{13}$. What is the length of side $AB$, rounded to the nearest integer if necessary?

---

## Problem 20

$z = -1 + i$ is one complex number such that

$$z^2 - z + 1 = -2iz - 5i.$$

There is one different complex number $z = a + bi$ that also satisfies this equation. For this complex number, what is $a^2 + b^2$?

## 1.7  ZIML April 2018 Division H

Below are the 20 Problems from the Division H ZIML Competition held in April 2018.
The answer key is available on p.178 in the Appendix.
Full solutions to these questions are available starting on p.130.

### Problem 1
Alice, Bob, and Eve are washing cars. In total they wash 600 cars. Alice washes a third as many cars as Bob and Eve combined. Eve washes half as many cars as Alice and Bob combined. How many cars does Bob wash?

### Problem 2
The circular sector with radius 10 shown below is folded together to make the lateral surface of a cone.

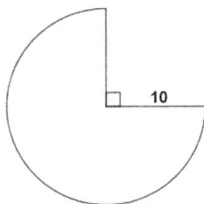

The area of the base of this cone can be written as $K \times \pi$. What is $K$? Round your answer to the nearest hundredth if necessary.

### Problem 3
There is one integer value of $C$ so that the graphs of $y = x + C$ and $y = 4 - 2|x+1|$ intersect at exactly one point. What is $C$?

## Problem 4

Randomly rearrange the 11 letters in the word PROBABILITY. The probability that the two B's and the two I's appear next to each other in this rearrangement can be expressed as $\dfrac{P}{Q}$ for positive integers $P, Q$ with $\gcd(P, Q) = 1$. What is $Q - P$?

## Problem 5

Let the integer $K$ be the smallest multiple of 896 and 1875 that is a perfect cube. How many zeros does $K$ end in?

## Problem 6

$\triangle ABC$ has sides $AB = 10$ and $AC = 8\sqrt{2}$. A circle is drawn containing $B$ and $C$ intersecting the triangle at $D$ and $E$ as shown in the diagram below.

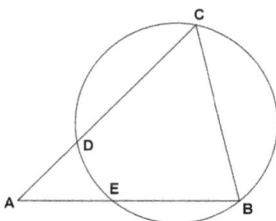

If the (minor) arcs have measures $\widehat{BC} = 130°$ and $\widehat{DE} = 40°$, what is $BC^2$? Round your answer to the nearest integer if necessary.

## Problem 7

Consider the real solutions to the equation

$$\log_8(x) + \log_x(8) = \frac{10}{3}$$

What is the difference between the largest solution and the smallest solution? Round your answer to the nearest tenth if necessary.

## Problem 8

Find all the solutions to $\sin^2(\theta) = 2\cos^2(2\theta) + \cos^2(\theta)$ for $0° \le \theta < 90°$ with $\theta$ an integer. What is the sum of all the solutions?

## Problem 9

Peter creates a solid by starting with a cube of side length 4 cm and adding a square pyramid with height 2 cm to the top, so that the base of the pyramid is is one of the faces of the original cube.

What is the volume of this solid? Round your answer to the nearest cubic centimeter if necessary.

## Problem 10

Solve the equation $(2x - 5)^3 + 45 = 18x$. What is the sum of all the integer solutions? (If there are no integer solutions, input an answer of 0.)

## Problem 11
What are the last two digits of $91^{91}$?

## Problem 12
Ariel is practicing her archery using a target made up of 4 circles. Each circle has a radius 1 unit more than the previous. She assigns point values to each region as shown below.

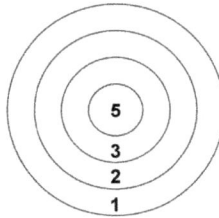

Ariel shoots two arrows with both arrows randomly landing somewhere on the target. The probability the two shots have a combined point total of 6 points can be expressed as $S\%$. What is $S$, rounded to the nearest integer if necessary?

## Problem 13

In the diagram below, there are 15 grid points arranged in equilateral triangles, equally spaced. The area of each small equilateral triangle formed by 3 adjacent grid points is 2.

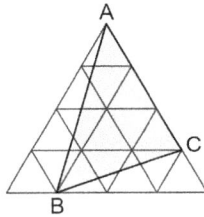

Find the area of $\triangle ABC$. Round your answer to the nearest integer if necessary.

## Problem 14

The expression $(\sqrt{3}+i)^{81}$ can be written as $2^K \cdot (A+Bi)$ for integers $K, A, B$, where $K$ is as large as possible.

What is $K + A + B$?

## Problem 15

The region inside the circle $x^2 - 6x + y^2 - 20y + 93 = 0$ that is below the line $y = 2x + 4$ has area $S \times \pi$. What is $S$, rounded to the nearest tenth if necessary?

## Problem 16

Consider $P(x) = 2x^3 - ax^2 - bx - 1$ where $a, b$ are integers. The remainder when $P(x) \div (x+1)$ is $-10$ and the remainder when $P(x) \div (x+2)$ is $-37$. What is $P(2)$?

## Problem 17

An octagon is created with 3 squares, 2 rectangles, and 4 30-60-90 triangles as shown in the diagram below.

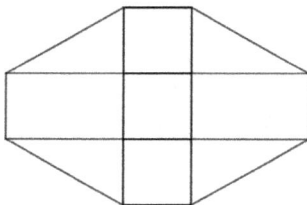

If the area of the octagon is $6 + 8\sqrt{3}$, what is the side length of the middle square? Round your answer to the nearest tenth if necessary.

## Problem 18

In $\triangle ABC$ and $\triangle DEF$, $\angle A = \angle D$ and $\angle B = \angle E$. If $AB = 30$, $BC = 15$, $DE = x+2$, and $EF = x-2$ for $x$ an integer, what is $x$?

## Problem 19

Izzy has a knapsack with 5 balls, numbered 1 through 5. The first 2 are yellow and the other 3 are purple. She randomly reaches into the knapsack and chooses one ball and then places it back. She does this a total of 5 times, recording the number of the ball each time. If Izzy picks 3 yellow and 2 purple balls, how many different outcomes are possible?

## Problem 20

If $T = \dfrac{6}{\sqrt{7}-1}$, what is $\lfloor T \rfloor + (2+\sqrt{7})\{T\}$? Round your answer to the nearest hundredth if necessary.

Here $\lfloor x \rfloor$ denotes the greatest integer $\leq x$ and $\{x\} = x - \lfloor x \rfloor$ and you may use the fact that $\sqrt{7} \approx 2.645751$.

## 1.8 ZIML May 2018 Division H

Below are the 20 Problems from the Division H ZIML Competition held in May 2018.
The answer key is available on p.179 in the Appendix.
Full solutions to these questions are available starting on p.140.

### Problem 1
Parallelogram $PQRS$ is shown below, where $T$ is the midpoint of $\overline{PQ}$.

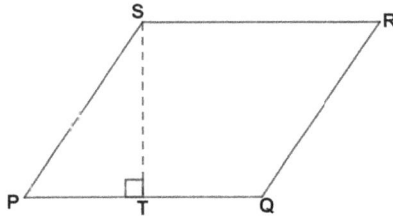

If $RS = 8$ and the area of $PQRS$ is 48, then the perimeter of $PQRS$ can be written as $A + B\sqrt{C}$ for positive integers $A$, $B$, and $C$ where $C$ has no factors that are squares. What is $A + B + C$?

### Problem 2
$x$ is a positive real number such that $\log_5(x^2 - 4) = 5$. What is $x$, rounded to the nearest integer?

## Problem 3

There are some toy cars in a box. 27 are collector's edition cars and 33 are regular toy cars. 23 of the cars are still in their unopened boxes and 37 of the cars do not have their box anymore. If we know at least $\frac{1}{3}$ of the collector's edition cars are in their unopened boxes, what is the least number of regular toy cars that do not have their boxes anymore?

## Problem 4

Find the sum of all real solutions of the equation $\sqrt{3x+2} + \sqrt{3x-2} = 3$. Round your answer to the nearest hundredth if necessary.

## Problem 5

A right cylinder of height 20 is inscribed in a a sphere of radius 26. What is the radius of the base of the cylinder? Round your answer to the nearest hundredth if necessary.

## Problem 6

Barry, Carrie, and Mary describe a number to Terry, who is supposed to guess the number. Barry says the number has 15 factors. Carrie says the number is a perfect square. Mary says the number is a multiple of 22. Terry states that the others must have been mistaken, because he found two numbers that match their description. What is the larger number minus the smaller number?

## Problem 7

How many real solutions does the equation

$$(x^3 + 4x^2 + x + 4)(x^3 - 8) = 0$$

have?

## Problem 8

Let $f(x) = x^2 - 1$. There is one integer $n$ such that $f(n+1) + f(n) = n$. What is this $n$?

## Problem 9

Consider the triangle shown below

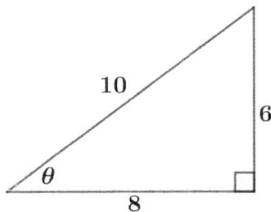

What is $\dfrac{1}{\sin(\theta)} - \dfrac{1}{\cos(\theta)}$, rounded to the nearest tenth if necessary.

## Problem 10

What is the minimum value achieved by the function

$$y = \frac{x^3 - 2x^2 - 5x + 6}{x + 2}$$

for $|x| \le 100$?

## Problem 11

Charlie goes to the candy store to buy some chocolate and some gum. The store offers 10 varieties of chocolate and 5 varieties of gum. Charlie buys three different types of chocolate and two different types of gum. How many different collections of candy (chocolate and gum) can Charlie buy? The order in which he buys the candy does not matter.

## Problem 12

The line $3y = x\sqrt{3} + 4\sqrt{3}$ intersects the circle $x^2 + y^2 = 16$ at two points. Consider the sector formed on the arc between these two points as shaded in the diagram below.

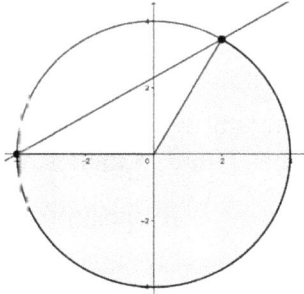

The area of this sector can be expressed as $\dfrac{P}{Q} \cdot \pi$, where $P$ and $Q$ are positive integers such that $\gcd(P, Q) = 1$. What is $P + Q$?

## Problem 13

If $90° < x < 180°$ and $6\cos^2(x) = 4 - \sin(x)$, what is $\sin(x)$? Round your answer to the nearest hundredth if necessary.

## Problem 14

Find the greatest common divisor of $5! + 6! + 7!$ and $7! + 8! + 9!$. Recall $N! = N \cdot (N-1) \cdot (N-2) \cdots 1$.

## Problem 15
A lattice point is a point $(x,y)$ in the plane such that $x$ and $y$ are both integers. How many lattice points lie on the line $x+3y=100$ in the first quadrant? Recall the first quadrant is all points such that $x>0$ and $y>0$.

## Problem 16
In triangle $\triangle ABC$ the angles are in ratio $3:4:5$. If the shortest side of the triangle has length 5 inches, the length of the middle side can be written as $\dfrac{P\sqrt{Q}}{R}$ for positive integers $P,Q,R$ such that $\gcd(P,R)=1$ and no squares are factors of $Q$. What is $P+Q+R$?

## Problem 17
Jason flips a fair coin 5 times. He records the outcomes using $H$ for heads and $T$ for tails. For example, one such outcome could be $HTHTH$. The probability Jason records an outcome that contains $HHH$ can be expressed as $\dfrac{P}{Q}$ for positive integers $P,Q$ with $\gcd(P,Q)=1$. What is $P+Q$?

## Problem 18
The graphs of the equations $y=|x|-6$ and $y=-|x-2|+2$ determine a rectangle. What is the area of this rectangle? Round your answer to the nearest integer if necessary.

## Problem 19

Find the sum of all real roots of

$$p(z) = z^4 - 12z^3 + 57z^2 - 126z + 104$$

given that $p(3 - 2i) = 0$.

## Problem 20

A cube with opposite edges $\overline{AB}$ and $\overline{CD}$ is shown below.

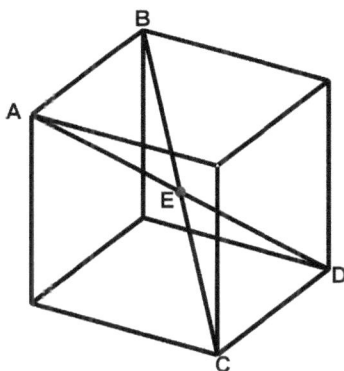

The diagonals $\overline{AD}$ and $\overline{BC}$ intersect at point $E$. What is $\cos(\angle AEB)$, rounded to the nearest hundredth if necessary?

## 1.9 ZIML June 2018 Division H

Below are the 20 Problems from the Division H ZIML Competition held in June 2018.
The answer key is available on p.180 in the Appendix.
Full solutions to these questions are available starting on p.149.

### Problem 1
Mario builds a model mushroom using a half-sphere and a cylinder as shown in the diagram below.

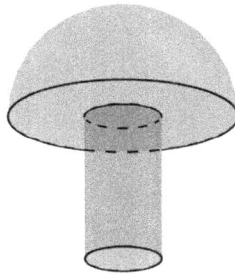

The height and diameter of the cylinder are in ratio $2 : 1$ while the ratio of the diameters of the cylinder and half-sphere is $1 : 3$. The ratio between the volume and surface area of the full model mushroom can be written as $P : Q$ for positive integers $P, Q$ with $\gcd(P, Q) = 1$. What is $P + Q$?

## Problem 2

August and Bernard want to exchange some books with each other. August has 6 books and Bernard has 4 books. August chose 2 of his books and gave them to Bernard. Bernard put the 2 books that August gave him in his book collection, and forgot which were his and which were August's. Then Bernard chose 2 books to give to August. In how many different ways could this book exchange have happened?

## Problem 3

Consider the function $f(x) = x^3 - 9x^2 + 27x - 23$ and $f^{-1}(x)$ its inverse. What is $f(23) - f^{-1}(-23)$ ?

## Problem 4

Charley is working with bacteria in his laboratory. If Charley places a sample of bacteria in an incubator, the number of bacteria doubles every 20 minutes. He places 50 samples of bacteria in the incubator, with each sample initially containing 200 bacteria, and checks on the samples every hour. How many hours does it take for there to be more than $5 \times 10^7$ bacteria in total?

## Problem 5

A ship travels at 25 mph with bearing $N\,15°E$ for 2 hours, then it travels at 20 mph with bearing $S\,75°W$ for 6 hours before coming to a full stop. The ship is now $\sqrt{m}$ miles away from its starting point. What is $m$? Round $m$ to the nearest integer if necessary.

Recall, for example $N\,15°E$, read as $15°$ east of north means the ship's path would form a $15°$ angle with due north and a $90° - 15° = 75°$ angle with due east.

---

## Problem 6

Find the sum of all real solutions to the equation

$$\log_3(x) + \log_9(x^2) + \log_{27}(x^3) = 9.$$

## Problem 7

Consider a convex pentagon $ABCDE$. Suppose $\angle EAC = \angle ABC$, $\angle BCD = 110°$, $\angle CDE = 102°$, and $\angle DEA = 170°$. What is the measure of the exterior angle at vertex $A$?

## Problem 8

How many perfect squares are factors of both $18000$ and $21600$?

## Problem 9

Yumi is playing a card game using a standard 52-card deck. During the game she is dealt 4 cards and she wins if he gets 4 cards of the same rank. In this game the cards of rank $A$ act as a wildcard, so an $A$ can be used as any other card. How many winning hands are there for this game?

Recall a standard deck of cards is 4 cards each for 13 ranks: $2, 3, \ldots, 10, J, Q, K, A$.

## Problem 10

Isabelle drew a circle of radius one on a piece of paper. Then she drew a second circle of radius one with its center on the first circle. Lastly she drew a third circle of radius one with its center on the second circle so that the area contained in all three circles was maximum. This overlapping area can be expressed as $\dfrac{A\pi + B\sqrt{3}}{C}$, where $A, B, C$ are integers with $C > 0$ and $\gcd(A, B, C) = 1$. What is $A + B + C$?

## Problem 11

Consider the function $f(x) = 3x^2 - 36x + 112$ with domain all $x \geq m$. Find the smallest value of $m$ such that $f(x)$ has an inverse.

## Problem 12

Consider the sequence defined by $G_1 = 3$, $G_2 = 5$, and $G_n = 3 \times G_{n-1} - G_{n-2}$ for $n \geq 3$. What is the remainder when $G_{2018}$ is divided by 7?

## Problem 13

In the diagram below, the large circle has radius 4.

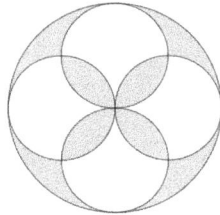

The shaded area is $A + B \times \pi$, where $A$ and $B$ are integers. What is $A + B$?

## Problem 14

What is the difference between the largest and the smallest solutions of

$$\cos(x) - \sin(x)\sin(2x) - \cos(2x) = 0,$$

where $0° \le x \le 180°$.

## Problem 15

The sum of 7 consecutive 3-digit positive integers is divisible by 154. If all the numbers are as small as possible, what is the smallest of the 5 numbers?

## Problem 16

Find the sum of all real solutions to the equation $|2x - |x+4|| = 16$. Round your answer to the nearest tenth if necessary.

## Problem 17
An equilateral triangle of side length 10 is divided into 100 triangles of side length 1 forming a triangular grid. 4 points on this grid are selected to form a parallelogram whose sides are parallel to the sides of the big triangle. How many different parallelograms can be formed?

## Problem 18
Calculate $\sum_{k=1}^{10}(-1)^k 2^{k-1}$.

## Problem 19
Justina just found out that a spider crawled on top of her Christmas tree and spun a long (straight) spiderweb all the way to the ground. The spiderweb makes an angle of $15°$ with the ground and her Christmas tree is 8 feet tall. The spiderweb is $8(\sqrt{A}+\sqrt{B})$ feet long, with $A$ and $B$ integers. What is $A \times B$?

Recall that $\sin(x-y) = \sin(x)\cos(y) - \cos(x)\sin(y)$.

## Problem 20
Consider the curves defined by $x^2 + 4y^2 - 16 = 0$ and $x^2 + y^2 + 2y = 0$. How many times do they intersect?

# 2. ZIML Solutions

This part of the book contains the official solutions to the problems from the nine Division H ZIML Contests from the 2017-18 School Year.

Students are encouraged to discuss and share their own methods to the problems using the Discussion Forum on ziml.areteem.org.

## 2.1  ZIML October 2017 Division H

Below are the solutions from the Division H ZIML Competition held in October 2017.
The problems from the contest are available on p.17.

### Problem 1 Solution
Of the 100 students, 80 played sports and 60 were on the honor roll. As this means $80 + 60 = 140$ awards were given in total, $140 - 100 = 40$ students must have gotten both awards.

Hence 40 of the 80 students who played sports were on the honor roll. This is $\frac{40}{80} = 50\%$, so $K = 50$.

**Answer:** 50

### Problem 2 Solution
Adding the two equations we get

$$\left(\frac{1}{2}\sqrt{x} + \frac{1}{2}\sqrt{x}\right) + (-2y + 2y) = 4 + 6$$

so $\sqrt{x} = 10$. Therefore we see $x = 100$. Substituting we get

$$\frac{\sqrt{100}}{2} - 2y = 4 \text{ so } 5 - 2y = 4$$

and therefore $-2y = -1$ so $y = 0.5$. Thus $A + B = 100 + 0.5 = 100.5$.

**Answer:** 100.5

### Problem 3 Solution
Note that, since $E$ and $F$ are midpoints, the area of $ABF$ and $BCE$ are both $1/4$ of the parallelogram. Hence each has area $20 \div 4 = 5$.

Similarly $DEF$ has area $1/8$ of the parallelogram, so has area $20 \div 8 = 2.5$. Therefore the remaining triangle $BEF$ has area $20 - 5 - 5 - 2.5 = 7.5$.

**Answer:** 7.5

### Problem 4 Solution

Note $\triangle APD$ is isosceles because $AD = AP$. Further

$$\angle PAD = 90° - \angle PAB = 90° - 60° = 30°.$$

Hence

$$\angle PDA = \angle DPA = (180° - 30°) \div 2 = 75°.$$

This gives

$$\angle CDP = 90° - \angle PDA = 90° - 75° = 15°$$

and thus 15 is our answer.

**Answer:** 15

### Problem 5 Solution

The arrow reaches its highest point at the vertex of the parabola. The line of symmetry of the parabola goes through the vertex, so it will take the same time for the arrow to get to the highest point and to go down to the launching height again.

Thus, it took $2.3 \times 2 = 4.6$ seconds for the arrow to reach the same height.

**Answer:** 4.6

### Problem 6 Solution

Since the sums of the two numbers on each card are equal, the back numbers of these three cards must each contain at least one even and one odd numbers. However, the back numbers are all primes, so one of them (the even one) has to be 2.

The only way it is possible is that the number on the back of the card with 51 on the front is 2. Therefore the sum is 53, and the other two hidden numbers are $53 - 12 = 41$ and $53 - 42 = 11$.

The average: $(41 + 2 + 11) \div 3 = 18$.

**Answer:** 18

## Problem 7 Solution

Note $200 \div 4 = 50$ of these numbers are divisible by 4. As 198 is the largest multiple of 6 less than or equal to 200, there are $198 \div 6 = 33$ numbers divisible by 6.

However, the multiples of 12 are being counted twice (once as a multiple of 4 and again as a multiple of 6). As there are $192 \div 12 = 16$ numbers divisible by 12 less than or equal to 200, in total we have that

$$50 + 33 - 16 = 67$$

of the first 200 positive integers are divisible by 4 or 6.

**Answer:** 67

## Problem 8 Solution

Since the equation has a root at $x = -1$ we know $(x + 1)$ is a factor. Similarly as $x = 2$ is a double root, $(x - 2)^2$ is a factor. Hence the equation must have the form $y = k(x + 1)(x - 2)^2$. For the $y$-intercept to be 8 we must have $8 = k \times 1 \times 4$ so $k = 2$. Therefore

$$y = 2(x + 1)(x - 2)^2 = 2x^3 - 6x^2 + 8$$

so $a + b + c + d = 2 - 6 + 0 + 8 = 4$.

**Answer:** 4

### Problem 9 Solution

Since $\angle ABC = 80°$, we have the measure of arc $AC$ (not containing $B$) is $2 \times 80° = 160°$. Hence, the measure of arc $ABC$ is $360° - 160° = 200°$.

As the ratio of the measures of arcs $AB : BC = 2 : 3$, we see their measures must be $200° \times \dfrac{2}{5} = 80°$ and $200° \times \dfrac{3}{5} = 120°$.

Thus, arc $BC$ has measure $120°$ so $\angle BAC = 120° \div 2 = 60°$.

**Answer:** 60

### Problem 10 Solution

If the line and parabola intersect, they intersect at $x + K = x^2 - 5x + 14$. Hence we need to ensure that $x^2 - 6x + (14 - K) = 0$ has real solutions. Using the discriminant we have

$$(-6)^2 - 4 \times 1 \times (14 - k) \geq 0$$
$$\Rightarrow 36 + 4k - 56 \geq 0$$
$$\Rightarrow 4k \geq 20$$
$$\Rightarrow K \geq 5.$$

Hence the line $y = x + K$ first intersects the parabola when $K = 5$.

**Answer:** 5

### Problem 11 Solution

We want $\sin(2\theta - 90°) = \dfrac{\sqrt{2}}{2}$. Recall

$$\sin(2\theta - 90°) = -\sin(90° - 2\theta) = -\cos(2\theta),$$

and therefore $\cos(2\theta) = -\dfrac{\sqrt{2}}{2}$.

Thus $2\theta = 135°, 225°, \ldots$, so

$$135° \div 2 = 67.5°.$$

is the smallest possible $\theta$.

**Answer:** 67.5

## Problem 12 Solution

Note any number 2 less than a multiple of 5 will leave a remainder of 3 when divided by 5. Similarly, any number 2 less than a multiple of 7 will leave a remainder of 5 when divided by 7 and any number 2 less than a multiple of 9 will leave a remainder of 7 when divided by 9.

Therefore, as 315 is the least common multiple of 5, 7, and 9, we see $315 - 2 = 313$ is the smallest number that will leave the correct remainders.

**Answer:** 313

## Problem 13 Solution

The circle intersects the line $x = 4$ when $(4-2)^2 + (y+2)^2 = 25$ so $(y+2)^2 = 21$ so $y = -2 \pm \sqrt{21}$. Hence if we view $\triangle ABC$ as having base $AB$ we have a base of length

$$-2 + \sqrt{21} - (-2 - \sqrt{21}) = 2\sqrt{21}$$

with height $4 - 2 = 2$. This triangle has area $2\sqrt{21}$, so $R + S = 2 + 21 = 23$.

**Answer:** 23

## Problem 14 Solution

We know the two roots are complex conjugates, so $p = q = 2$. Hence the roots are $x = 2 \pm i\sqrt{2}$. As the coefficient of $x^2$ is 3 the quadratic must be

$$3(x-(2+i\sqrt{2}))(x-(2-i\sqrt{2}))$$
$$= 3(x^2 - 4x + 6)$$
$$= 3x^2 - 12x + 18.$$

Therefore $C = 18$.

**Answer:** 18

## Problem 15 Solution

We know $\log_3 1 = 0$, so we want $\log_2 |x^2 - 9| = 1$. As $\log_2 2 = 1$, we must have $|x^2 - 9| = 2$. This implies $x^2 - 9 = \pm 2$ so $x^2 = 7$ or $x^2 = 11$.

Hence the roots are $\pm\sqrt{7}, \pm\sqrt{11}$ so $K + L = 7 + 11 = 18$.

**Answer:** 18

## Problem 16 Solution

Note we can rewrite the left-hand side as $2^x(3^x + 4^x)$.

Since $3^x$ is always odd and $4^x$ is always even, we know $3^x + 4^x$ is always odd.

Hence when we factor $5392 = 2^4 \times 337$, we see that $2^x = 2^4$ and $x$ must be 4. Double checking, $3^4 + 4^4 = 81 + 256 = 337$ as needed.

**Answer:** 4

## Problem 17 Solution

We know all the sides for $\triangle ABD$, so using the law of cosines we have $AD^2 = AB^2 + BD^2 - 2 \times AB \times BD \times \cos \angle ABD$ so

$$8^2 = 6^2 + 4^2 - 2 \times 6 \times 4 \times \cos \angle ABD$$
$$\Rightarrow 12 = -48 \cos \angle ABD$$
$$\Rightarrow \cos \angle ABD = -\frac{1}{4}.$$

Therefore $\cos \angle CBD = \cos(180° - \angle ABD) = \frac{1}{4}$. Using the law of cosines for triangle $BCD$ then gives

$$CD^2 = BD^2 + BC^2 - 2 \times BD \times BC$$
$$= 4^2 + 2^2 - 2 \times 4 \times 2 \times \frac{1}{4}$$
$$= 16.$$

Hence $CD = \sqrt{16} = 4$.

**Answer:** 5

## Problem 18 Solution

Draw the altitude of the isosceles triangle as shown below.

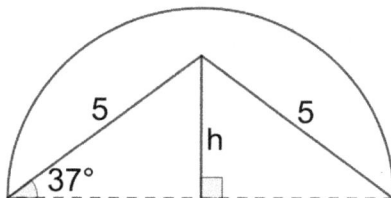

Therefore, $\frac{h}{5} = \sin(37°) \approx 0.6$ so $h = 3$. This means the radius of the semicircle is $\sqrt{5^2 - 3^2} = 4$.

Hence the semicircle has area $\frac{1}{2}\pi 4^2 = 8\pi \approx 24.8$. The removed triangle has area $\frac{8 \cdot 3}{2} = 12$, so the shaded region has an approximate area of $24.8 - 12 = 12.8$.

**Answer:** 12.8

### Problem 19 Solution

Using polynomial long division we have $(x^4 - 2x^3 - 9x^2 - 2x + 8) \div (x^2 - 3x - 5) = (x^2 + x - 1)$ with remainder 3, so we have

$$x^4 - 2x^3 - 9x^2 - 2x + 8$$
$$= (x^2 - 3x - 5)(x^2 + x - 1) + 3$$
$$= 0 + 3 = 3$$

using the quotient and remainder when we divide.

**Answer:** 3

### Problem 20 Solution

There are 8 students in total, so there are $8 \times 7 = 56$ total ways to choose the two leaders.

We either want a girl and then a boy to be chosen, $3 \times 5 = 15$ ways, or a boy and then a girl to be chosen, $5 \times 3 = 15$ ways.

Hence there is a $\frac{30}{56} = \frac{15}{28}$ chance that one girl and one boy are picked. Thus $P + Q = 15 + 28 = 43$.

**Answer:** 43

## 2.2   ZIML November 2017 Division H

Below are the solutions from the Division H ZIML Competition held in November 2017.
The problems from the contest are available on p.23.

### Problem 1 Solution

We know $\overline{AB} \parallel \overline{CD}$ because $ABCD$ is a trapezoid. Therefore $\triangle ABE$ is similar to $\triangle DCE$ as they share the same three angles. Let $DE = x$, so

$$\frac{DE}{AE} = \frac{CD}{AB} \Rightarrow \frac{x}{x+3} = \frac{4}{8}.$$

Solving for $x$ we have $2x = x+3$ so $x = 3$. An identical calculation shows $CE = 2$.

Therefore the perimeter of $\triangle CDE$ is $2+3+4 = 9$.

**Answer:** 9

### Problem 2 Solution

Checking possible roots from the Rational Root Theorem, we see $-1, 2$ are roots, so we can divide by $(x+1)(x-2) = x^2 - x - 2$. Using long division this gives us

$$(x^4 - 7x^3 + x^2 + 15x + 6) \div (x^2 - x - 2) = x^2 - 6x - 3,$$

so the other roots are $3 \pm \sqrt{12}$ using the quadratic formula.

Hence $A + B = 3 + 12 = 15$.

**Answer:** 15

### Problem 3 Solution

Pretend that all of the letters are distinct. There are

$$7! = 7 \times 6 \times 5 \times 4 \times 3 \times 2 \times 1 = 840$$

ways to rearrange 7 letters to form distinct words.

Since there are 3 $A$'s and 2 $N$'s in *BANANAS*, we need to rid the duplicates by dividing the total number of ways to rearrange 7 different letters by $3! \times 2! = 6 \times 2 = 12$.

Therefore, the answer is

$$\frac{7!}{3! \cdot 2!} = \frac{7 \times 6 \times 5 \times 4 \times 3 \times 2 \times 1}{3 \times 2 \times 1 \times 2 \times 1} = 420.$$

**Answer:** 420

## Problem 4 Solution
The vertex occurs when

$$x = \frac{-b}{2a} = \frac{-72}{2 \times -36} = 1.$$

Since the $x^2$ coefficient is negative, $f(x)$ has a maximum. Plugging in $x = 1$ we have $f(1) = 97$ is the global maximum.

**Answer:** 97

## Problem 5 Solution
The region can be divided into one big sector of $300°$ (5/6 of a circle with radius 4) and two small sectors of $120°$ (each is 1/3 of a circle with radius 1). Thus the area is

$$\frac{5}{6} \cdot 4^2 \pi + \frac{2}{3} \cdot 1^2 \pi = 14\pi,$$

so $K = 14$.

**Answer:** 14

### Problem 6 Solution

We look for a pattern in $6^1, 6^2, 6^3, \ldots$. Note that in the pattern only the ones and tens digit matter as we only care about the tens digit in the end. We have

| $N:$ | 1 | 2 | 3 | 4 | 5 | 6 | 7 |
|---|---|---|---|---|---|---|---|
| Last two digits of $6^N:$ | 06 | 36 | 16 | 96 | 76 | 56 | 36 |

Note from here the pattern will continue

$$36, 16, 96, 76, 56, \ldots$$

repeating every 5 terms.

Hence $6^3, 6^8, 6^{13}, 6^{18}, \ldots, 6^{2018}$ will all have last two digits 16 (as $2018 = 403 \times 5 + 3$). Therefore the tens digit is 1.

**Answer:** 1

### Problem 7 Solution

First note the sector itself has a measure of $360° - 80° = 280°$. Hence the arc length of the sector is given by

$$\frac{280°}{360°} \times 2 \times \pi \times 6 \approx \frac{7}{9} \times 36 = 28.$$

For the perimeter we also need the two radii, which add another $6 + 6 = 12$ to the perimeter.

The total perimeter is thus $28 + 12 = 40$ units.

**Answer:** 40

### Problem 8 Solution

Rewrite the equation as $x^2 + 3x + (6 - m) = 0$. The discriminant is

$$b^2 - 4ac = 3^2 - 4 \times 1 \times (6 - m)$$
$$= 9 - 24 + 4m$$
$$= -15 + 4m.$$

For exactly one solution, we must have $4m - 15 = 0$ or $m = \dfrac{15}{4}$. Hence as a decimal, $m = 3.75$.

**Answer:** 3.75

## Problem 9 Solution

Larry only needs one more heads to win the bet. Therefore the only way to lose is if the next 3 flips are all tails. This has probability

$$\frac{1}{2} \times \frac{1}{2} \times \frac{1}{2} = \frac{1}{8} = 0.125 = 12.5\%.$$

Hence Larry has a 12.5% chance of losing his bet, so has a $100\% - 12.5\% = 87.5\% \approx 88\%$ chance to win his bet. Thus $K = 88$.

**Answer:** 88

## Problem 10 Solution

Let's examine $17x - x^2 = x(17 - x)$.

The graph of $y = x(17 - x)$ is a parabola opening downwards with zeros at $x = 0$ and $x = -17$. Hence $y = x(17 - x)$ is less than zero if $x < 0$ or $x > 17$.

This implies the domain of $y = \sqrt{17x - x^2}$ is all $x$ with $0 \le x \le 17$. This is a total of $17 - 0 + 1 = 18$ integers.

**Answer:** 18

## Problem 11 Solution

Note the stick will rest on opposite sides of the circle on the base and at the top, giving a side-view as shown below.

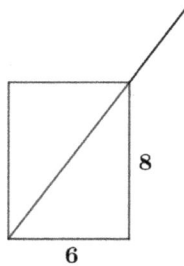

Since the radius of the base is 3 cm, the diameter is 6 cm. Hence using the Pythagorean theorem, the length of the stick inside the cup is $\sqrt{6^2 + 8^2} = \sqrt{100} = 10$ cm. Hence there are $15 - 10 = 5$ cm that stick out of the cup.

**Answer:** 5

## Problem 12 Solution

Rewrite the equation to

$$2 \lfloor x \rfloor = x + \{x\} = \lfloor x \rfloor + \{x\} + \{x\},$$

so $\lfloor x \rfloor = 2\{x\}$. Since $0 \le \{x\} < 1$, and $\lfloor x \rfloor$ is an integer, we get $\{x\} = 1/2$, and $\lfloor x \rfloor = 1$.

Therefore $x = 1 + 1/2 = 3/2 = 1.5$.

**Answer:** 1.5

### Problem 13 Solution

We have (using change of base)

$$\log_3(x) + \log_9(x) + \log_{27}(x)$$
$$= \log_3(x) + \frac{\log_3(x)}{\log_3(9)} + \frac{\log_3(x)}{\log_3(27)}$$
$$= \log_3(x) + \frac{\log_3(x)}{2} + \frac{\log_3(x)}{3}$$
$$= \frac{11\log_3(x)}{6}.$$

Hence we need $\log_3(x) = \frac{22}{3} \cdot \frac{6}{11} = 4$ so $x = 3^4 = 81$.

**Answer:** 81

### Problem 14 Solution

We have

$$\frac{5}{\sin(30°)} = \frac{AC}{\sin(45°)}$$

or, after rearranging,

$$AC = 5 \cdot \frac{\sin(45°)}{\sin(30°)} = 5 \cdot \frac{\sqrt{2}/2}{1/2} = 5\sqrt{2} \approx 7.07.$$

Rounded to the nearest tenth we have $AC \approx 7.1$.

**Answer:** 7.1

### Problem 15 Solution

The smallest possible median is 0, which happens if at least 5 of the elements are 0.

We then make the remaining elements as large as possible (each 20), giving a mean of

$$(5 \times 0 + 4 \times 20) \div 9 = \frac{80}{9} \approx 8.8889 \approx 8.9.$$

This gives a difference of $\approx 8.9 - 0 = 8.9$, the largest possible.

Note it is also possible to achieve this difference by making the median as large as possible (20) and make the remaining elements 0. In this case the median is 20 and the mean is $\approx 11.1$, so the difference is still $\approx 8.9$.

**Answer:** 8.9

## Problem 16 Solution

Substituting $\sin^2(\theta) = 1 - \cos^2(\theta)$ we can rewrite the whole equation in terms of cosine and factor:

$$3\cos(\theta) = 2 - 2\cos^2(\theta)$$
$$\Rightarrow 2\cos^2(\theta) + 3\cos(\theta) - 2 = 0$$
$$\Rightarrow (2\cos(\theta) - 1)(\cos(\theta) + 2) = 0$$

Hence either $\cos(\theta) = \dfrac{1}{2}$ or $\cos(\theta) = -2$. Since the second case is impossible we must have $\cos(\theta) = \dfrac{1}{2}$. $\theta = 60°$ is the only solution in the range $0° \le \theta \le 90°$.

**Answer:** 60

## Problem 17 Solution

Note $\overline{EF}, \overline{GH}$ divide $ABCD$ into four parallelograms, whose areas are proportional. That is

$$\frac{[AEPG]}{[GPFD]} = \frac{[EBHP]}{[PHCF]} \Rightarrow [AEPG] = 12 \times \frac{20}{8} = 30.$$

Hence the total area of $ABCD$ is $30 + 20 + 8 + 12 = 70$.

**Answer:** 70

**Problem 18 Solution**

We know that $2^{11}$ has $11 + 1 = 12$ factors $(2^0, 2^1, \ldots, 2^{11})$.

Pairing the factors: $2^0$ with $2^{11}$, $2^1$ with $2^{10}$, etc., we see that each of the $12 \div 2 = 6$ pairs multiplies to $2^{11}$.

Hence the product of all the factors is $(2^{11})^6 = 2^{66}$ so $M = 66$.

**Answer:** 66

**Problem 19 Solution**

Starting to look for a pattern we have

$$(1 - i\sqrt{3})^2 = 1 - 2i\sqrt{3} - 3 = -2 - 2i\sqrt{3}.$$

Hence

$$(1 - i\sqrt{3})^3 = (1 - i\sqrt{3})(-2 - 2i\sqrt{3})$$
$$= -2 - 2i\sqrt{3} + 2i\sqrt{3} - 6 = -8.$$

As $10 = 3 \times 3 + 1$ we have

$$(1 - i\sqrt{3})^{10} = [(1 - i\sqrt{3})^3]^3 \times (1 - i\sqrt{3})$$
$$= (-8)^3 \times (1 - i\sqrt{3})$$
$$= -512(1 - i\sqrt{3})$$

so $K = -512$.

**Answer:** $-512$

**Problem 20 Solution**

Consider triangle $\triangle ECF$. Since we constructed equilateral triangles we know $EC = BC = 2$ and $CF = CD = 1$. Further we have $\angle ECF = 360° - 60° - 90° - 60° = 150°$. Using the law of cosines,

$$EF^2 = EC^2 + CF^2 - 2 \times EC \times CF \times \cos \angle ECF$$
$$= 2^2 + 1^2 - 2 \times 2 \times 1 \times \cos 150°$$
$$= 5 + 4 \times \frac{\sqrt{3}}{2} = 5 + 2\sqrt{3}.$$

As $\sqrt{3} \approx 1.73$ we have $EF^2 = 5 + 2\sqrt{3} \approx 5 + 3.46 = 8.46 \approx 8$.

**Answer:** 8

## 2.3 ZIML December 2017 Division H

Below are the solutions from the Division H ZIML Competition held in December 2017.
The problems from the contest are available on p.29.

### Problem 1 Solution
Note with a common denominator we have

$$\frac{x+3}{x-3} - \frac{x+2}{x-2} = \frac{(x^2+x-6)-(x^2-x-6)}{(x-3)(x-2)} = \frac{2x}{x^2-5x+6}.$$

Hence after cross-multiplying, $2x = x^2 - 5x + 6$ so

$$x^2 - 7x + 6 = (x-6)(x-1) = 0$$

and $x = 6$ or $x = 1$. Both solutions work. Thus the sum of the solution is $1 + 6 = 7$.

**Answer:** 7

### Problem 2 Solution
Label the couples $A$, $B$, $C$, and $D$. Since each couple wants to sit as a pair, first arrange the 4 couples, which can be done in

$$4! = 4 \times 3 \times 2 \times 1 = 24 \text{ ways.}$$

Then note that each of the couples themselves can be seated in 2 ways (Adam then Alice or Alice then Adam). This gives a total of
$$4! \times 2^4 = 24 \times 16 = 384$$

seating arrangements.

**Answer:** 384

## Problem 3 Solution

We in fact have that $\triangle ABO, \triangle BCO$ are both equilateral with side length 4.

Hence the area of the rhombus is twice the area of an equilateral triangle with side 4, so the area is

$$2 \times \frac{4^2\sqrt{3}}{4} = 8\sqrt{3}.$$

As $1.7 < \sqrt{3} < 1.8$ we have the area is between 13.6 and 14.4 so rounded to the nearest integer the area is 14.

**Answer:** 14

## Problem 4 Solution

Note
$$g(u) = u^2 - 13u + 12 = (u-1)(u-12)$$
is a parabola opening upwards with roots at $u = 1, 12$. Thus $g(u)$ is negative for $1 \leq u \leq 12$.

Substituting $u = x^2$ we have $f(x)$ is negative when $1 \leq x^2 \leq 12$ or $1 \leq |x| \leq 2\sqrt{3}$.

$2\sqrt{3} \approx 3.46$, thus $f(n)$ is negative for $n = \pm1, \pm2, \pm3$, so the answer is 6.

**Answer:** 6

## Problem 5 Solution

Recall the shortest distance is perpendicular to the line. Hence we find where $y = -3x - 10$ intersects the line with slope $m = -\frac{1}{3}$ containing the point $(0,0)$, which is

$$y = -\frac{1}{3}x.$$

These two lines intersect at the point $(3, -1)$. Hence the distance is

$$\sqrt{(3-0)^2 + (-1-0)^2} = \sqrt{10},$$

so $D = 10$.

**Answer:** 10

### Problem 6 Solution

Since the checkers are in different rows and columns, the 4 checkers will be placed in the 'top', 'bottom', 'left', and 'right' edges of the chessboard. Further, none of the 4 corners is possible, so we only have to ensure the 'top' and 'bottom' checkers and the 'left' and 'right' checkers do not attack each other.

There are 6 columns to choose from for the top checker, which leaves 5 columns for the bottom checker. Hence there are $6 \cdot 5 = 30$ ways to place the top and bottom checkers.

Similarly there are 30 ways for the left and right checkers, which leads to $30^2 = 900$ total ways.

**Answer:** 900

### Problem 7 Solution

Zeros at the of a number come from powers of 10 as factors. Since $10 = 2 \times 5$ we need to know how many powers of 2 and powers of 5 occur in the prime factorization of 200!. In fact, as there are clearly more powers of 2 than powers of 5, we just need to count how many powers of 5 occur.

Multiples of $5^1 = 5$: $5, 10, 15, 20, \ldots, 200$, a total of $200 \div 5 = 40$ numbers. Each contributes one power of 5.

Multiples of $5^2 = 25$: $25, 50, \ldots, 200$, a total of $200 \div 25 = 8$ numbers. Each contributes one extra power of 5.

Lastly, $5^3 = 125$ contributes one more power of 5.

In total there are
$$40 + 8 + 1 = 49$$
powers of 5 in the prime factorization of 200!, so there are 49 zeros at the end of 200!.

**Answer:** 49

**Problem 8 Solution**
Using the distance formula, we calculate the radius of the circle is
$$\sqrt{(5-3)^2 + (-2-4)^2} = \sqrt{4+36} = \sqrt{40}.$$
Thus the equation of the circle is
$$(x-3)^2 + (y-4)^2 = 40.$$
Expanding and rearranging we get
$$x^2 + y^2 = 6x + 8y + 15,$$
so $C = 15$.

**Answer:** 15

**Problem 9 Solution**
Using change of base formula, we have
$$\log_3(x) + 2\log_9(x) + 3\log_{27}$$
$$= \log_3(x) + \frac{2\log_3(x)}{\log_3(9)} + \frac{3\log_3(x)}{\log_3(27)}$$
$$= \log_3(x) + \frac{2\log_3(x)}{2} + \frac{3\log_3(x)}{3}$$
$$= 3\log_3(x)$$
Hence $\log_3(x) = 3$, so $x = 3^3 = 27$.

**Answer:** 27

## Problem 10 Solution

Since $O$ and $Q$ are on the $x$-axis, if $OP = PQ$ then $P$ has $x$-coordinate

$$\frac{0 + 2\sqrt{2}}{2} = \sqrt{2}.$$

Plugging this into the parabola we find $P = (\sqrt{2}, 2)$ and thus $OPQ$ is a triangle with base $OQ = 2\sqrt{2}$ and height 2.

Hence the triangle has area $2\sqrt{2} = \sqrt{8}$ so our answer is 8.

**Answer:** 8

## Problem 11 Solution

In total there are $6 \times 6 = 36$ outcomes.

Peter's roll is one of $1, 2, 3, 4, 5, 6$. For each of these cases there are, respectively, $1, 2, 3, 4, 5, 6$ possible rolls for Paul that are less than or equal to Peter's (for example if Peter rolls a 4, Paul could roll $1, 2, 3, 4$).

Thus the probability is

$$\frac{1 + 2 - 3 + 4 + 5 + 6}{36} = \frac{21}{36} = \frac{7}{12}$$

so $P + Q = 7 + 12 = 19$.

**Answer:** 19

## Problem 12 Solution

We know that $|2x + 1| = 2x + 1$ if $x \geq -\frac{1}{2}$ and $|2x + 1| = -2x - 1$ if $x < -0.5$.

First assume $x \geq -\frac{1}{2}$ so we have

$$|x - 2x - 1| = 3 \text{ or } |-x - 1| = 3.$$

---

This is the same as $|x+1| = 3$. If $x+1 = 3$ we have $x = 2$ as a solution and if $x+1 = -3$ we get $x = -4$. However, remember we are assuming $x \geq -\frac{1}{2}$ for this case, so $x = -4$ does not work. Hence $x = 2$ is the only solution with $x \geq -\frac{1}{2}$.

If $x < -\frac{1}{2}$ we have

$$|x + 2x + 1| = 3 \text{ or } |3x + 1| = 3.$$

If $3x + 1 = 3$ we get $x = \frac{2}{3}$. However, this is not less than $-\frac{1}{2}$, so it does not work. Else we have $3x + 1 = -3$ or $x = -\frac{4}{3}$.

Therefore $x = -\frac{4}{3}$ and $x = 2$ are the solutions. Only $\frac{-4}{3}$ is of the correct form with $Q - P = 3 - (-4) = 7$.

**Answer:** 7

### Problem 13 Solution
We have
$$\frac{\sin(30°)}{5} = \frac{\sin(45°)}{b}$$
or, after rearranging,
$$b = 5 \cdot \frac{\sin(45°)}{\sin(30°)} = 5 \cdot \frac{\sqrt{2}/2}{1/2} = 5\sqrt{2} \approx 7.07.$$
Therefore side $B \approx 7.1$ when rounded to the nearest tenth.

**Answer:** 7.1

### Problem 14 Solution
Let $u = \sin(x)$, so our equation becomes
$$4u^2 + u - 3 = 0.$$

Factoring we have

$$(4u - 3)(u + 1) = 0$$

so $u = \dfrac{3}{4}$ or $u = -1$.

If $\sin(x) = -1$, then $x$ is not between $0°$ and $90°$, hence $\sin(x) = u = \dfrac{3}{4}$. Using $\sin^2(x) + \cos^2(x) = 1$ we have

$$\cos^2(x) = 1 - \left(\frac{3}{4}\right)^2 = 1 - \frac{9}{16} = \frac{7}{16} = 0.4375.$$

Rounding we have an answer of 0.44.

**Answer:** 0.44

### Problem 15 Solution
The number 1 has 1 factor. Any prime number has 2 factors (1 and itself). Any number of the form $p^2$ for a prime $p$ has exactly 3 factors (1, $p$, and $p^2$). Any other number has at least 4 factors.

Since $22^2 = 484$ and $23^2 = 529$ we must find all primes $\leq 22$. They are 2, 3, 5, 7, 11, 13, 17, and 19, so there are 8 numbers $\leq 500$ with exactly 3 factors.

**Answer:** 8

### Problem 16 Solution
Let $g(x) = -cx^3$ so that $f(x) = g(x) + 20$.

We then know $g(3) = f(3) - 20 = 15 - 20 = -5$. Further, $g(x)$ is an odd function, so $g(x) = -g(-x)$ for all $x$. Therefore $g(-3) = -g(3) = -(-5) = 5$.

Thus $f(-3) = 5 - 20 = 25$.

**Answer:** 25

## Problem 17 Solution

Let $E$ and $F$ be points on $AB$ such that $DC \perp AB$ and $CD \perp AB$, forming a rectangle and two right triangles as shown below.

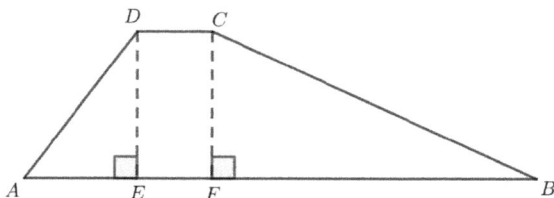

As $\sin A = \dfrac{4}{5}$, we have $DE = 12$ so using the Pythagorean theorem (or Pythagorean triples) $AE = 9$.

Thus $FB = 31 - 6 - 9 = 16$ and again using the Pythagorean theorem $CB = 20$. Therefore, the perimeter of the trapezoid is $15 + 6 + 20 + 31 = 72$.

**Answer:** 72

## Problem 18 Solution

$i^2 = -1$, so $i^3 = -i$, $i^4 = 1$, and $i^5 = i$. We have

$$f(i) = 6i^5 + 5i^4 + 4i^3 + 3i^2 + 2i + 1$$
$$= 6i + 5 - 4i - 3 + 2i + 1$$
$$= 3 + 4i$$

so $A^2 + B^2 = 3^2 + 4^2 = 25$.

**Answer:** 25

## Problem 19 Solution

Let $A = (0,0)$ and $B = (0,-2)$, so $A$ is on $y = mx$ and $B$ is on $y = x - 2$.

Let $C$ be the intersection of $y = mx$ and $y = x - 2$ (so $x_C > 0$)and consider triangle $ABC$.

We are given that $\angle B = 45°$. We also know that

$$\angle A = 20° + 90° = 110°$$

since the $x$ and $y$ axes are perpendicular. Therefore the final angle is

$$\angle C = 180° - 45° - 110° = 25°$$

which is the acute angle formed when the two lines meet and $K = 25$.

**Answer:** 25

**Problem 20 Solution**

The number of dollars is a multiple of $\text{lcm}(44, 77) = 308$, so there are $308k$ dollars, for some positive integer $k$.

Based on the description, there must be $308k \div 44 = 7k$ nieces and $308k \div 77 = 4k$ nephews.

It follows that each niece or nephew gets $308k \div (7k + 4k) = 308 \div 11 = 28$ dollars.

**Answer:** 28

## 2.4   ZIML January 2018 Division H

Below are the solutions from the Division H ZIML Competition held in January 2018.
The problems from the contest are available on p.35.

### Problem 1 Solution

Each year the car is worth 20% less than the year before, or equivalently it is worth 80% of the previous year's price. Hence after 3 years the car is worth

$$15000 \times 0.8^3 = 15000 \times 0.512 = 7680$$

dollars.

**Answer:** 7680

### Problem 2 Solution

The quadratic equation $3x^2 + bx + 12 = 0$ has no real solutions when the discriminant is negative:

$$b^2 - 4 \cdot 3 \cdot 12 < 0 \Rightarrow b^2 < 144 \Rightarrow -12 < b < 12.$$

Hence the integers are $-11, -10, \ldots, 10, 11$, a total of 23 integers.

**Answer:** 23

### Problem 3 Solution

We have

$$\frac{4+i}{3-i} = \frac{(4+i)(3+i)}{(3-i)(3+i)} = \frac{12+7i+i^2}{9+1} = \frac{11}{10} + \frac{7}{10}i.$$

Therefore, the imaginary part is $\dfrac{7}{10} = 0.7$.

**Answer:** 0.7

## Problem 4 Solution

$D$ must divide

$$571 - 513 = 58 = 2 \times 29$$
$$\text{and } 658 - 571 = 87 = 3 \times 29.$$

Therefore $D$ divides $\gcd(58, 87) = 29$. As 29 is prime, it's only factors are 1 and 29, so as $D > 1$ we must have $D = 29$.

Then note $513 \div 29 = 17 \times 29 + 20$ so the remainder $R = 20$.

**Answer:** 20

## Problem 5 Solution

Since the triangles are similar, their sides are proportional. Thus

$$\frac{20}{x+2} = \frac{12}{x-2} \Rightarrow 20x - 40 = 12x + 24.$$

Solving we have $x = 8$.

**Answer:** 8

## Problem 6 Solution

There are 7 different letters in total (H, A, P, Y, N, E, W, R). Hence there are $\binom{8}{2} = \frac{8 \times 7}{2} = 28$ ways to choose 2 different letters (in no particular order).

4 of these letters (A, P, Y, E) are repeated more than once, so it is possible for Peter to grab two sheets of the same letter repeated twice.

This gives 4 additional collections of sheets, so there are $28 + 4 = 32$ total collections.

**Answer:** 32

## Problem 7 Solution

We first calculate the units digits of $19^{19}$ and $99^{99}$. The pattern of the units digit is: $9, 1, 9, 1, \ldots$ for both.

The exponents are both odd, thus the units digits are each 9, and the units digit of the sum is 8.

**Answer:** 8

## Problem 8 Solution

Since $\overline{AB} \| \overline{CD}$, $\triangle ABE \sim \triangle CDE$, with ratio of corresponding sides $3 : 5$. Hence $DE : EB = 5 : 3$ and since $\triangle AED, \triangle AEB$ share the same height from $A$, $[AED] : [AEB] = 5 : 3$ so

$$[AEB] = 15 \cdot \frac{3}{5} = 9$$

We can use a similar argument to get

$$[DEC] = 15 \cdot \frac{5}{3} = 25$$

and lastly $[BEC] = 15$. Hence

$$15 + 9 + 25 + 15 = 64$$

is the total area.

**Answer:** 64

## Problem 9 Solution

The shortest distance satisfying the above conditions can be determined as follows:

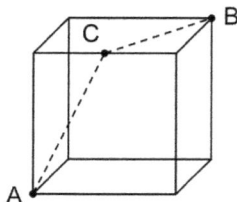

Note that the length of $AC$ and $CB$ can be determined by applying Pythagorean theorem on a right triangle with leg lengths 6 and 3. Therefore,

$$AC = BC = \sqrt{6^2 + 3^2} = \sqrt{45}$$

and the length of the shortest path from point $A$ to $B$ is $2\sqrt{45} = \sqrt{180}$, so $K = 180$.

**Answer:** 180

## Problem 10 Solution

We know $x = 3$ is a solution, so $(x - 3)$ is a factor of $x^3 - 5x^2 + 5x + 3$. Using long division we have

$$x^3 - 5x^2 + 5x + 3 = (x - 3)(x^2 - 2x - 1) = 0.$$

Using the quadratic formula $x^2 - 2x - 1 = 0$ has roots

$$\frac{2 \pm \sqrt{4 - 4 \times -1}}{2} = \frac{2 \pm 2\sqrt{2}}{2} = 1 \pm \sqrt{2}.$$

Therefore, $A + B = 1 - 2 = 3$.

**Answer:** 3

## Problem 11 Solution

We have $4000 = 2^5 \times 5^3$. Therefore the perfect squares are

$$2^0 \times 5^0 = 1, \quad 2^2 \times 5^0 = 4, \quad 2^4 \times 5^0 = 16,$$
$$2^0 \times 5^2 = 25, \quad 2^2 \times 5^2 = 100, \quad 2^4 \times 5^2 = 400,$$

all the factors with even exponents.

Thus there are 6 total factors that are also perfect squares.

**Answer:** 6

### Problem 12 Solution

Let $d$ be the distance to the flag. The ground you walked on and the flag form a right triangle, and we have $\cos(30°) = \dfrac{15}{d}$. This gives

$$d = \frac{15}{\sqrt{3}/2} = \frac{30}{\sqrt{3}} = 10\sqrt{3} \approx 17.3$$

as $\sqrt{3} \approx 1.73$. Rounded to the nearest integer you are 17 feet from the flag.

**Answer:** 17

### Problem 13 Solution

Of the 1000 shapes, $\dfrac{200}{1000} = \dfrac{1}{5}$ are squares and the other $\dfrac{4}{5}$ are not squares.

For the $\dfrac{1}{5}$ that are squares, the program (correctly) identifies them as a square 95% of the time, so

$$\frac{1}{5} \times 95\% = 19\%$$

of the time the program will (correctly) identify a shape as a square.

For the $\dfrac{4}{5}$ that are not squares, the program (incorrectly) identifies them as a square $100\% - 90\% = 10\%$ of the time, so

$$\frac{4}{5} \times 10\% = 8\%$$

of the time the program will (incorrectly) identify a shape as a square.

In total, $19\% + 8\% = 27\%$ of the time a shape is identified as a square.

**Answer:** 27

---

## Problem 14 Solution

Note the general term is of the form

$$\frac{1}{\sqrt{n}+\sqrt{n+1}} = \frac{\sqrt{n}-\sqrt{n+1}}{n-(n+1)} = \sqrt{n+1}-\sqrt{n}.$$

Hence the sequence becomes

$$\sqrt{2}-\sqrt{1}+\sqrt{3}-\sqrt{2}+\sqrt{4}-\sqrt{3}+\cdots+\sqrt{100}-\sqrt{99}$$

which after canceling is $\sqrt{100}-\sqrt{1} = 10-1 = 9.$

**Answer:** 9

## Problem 15 Solution

Adding the two equations we have

$$(2|x|+|x|)+(y-y) = 5+1$$

so $3|x| = 6$ or $|x| = 2$. Hence $x = 2$ or $x = -2$.

In either case we have $2(2)+y = 5$ so $y = 1$.

Hence our solutions are $(2,1)$ and $(-2,1)$. The product $x \cdot y$ is negative only for $(-2,1)$, giving an answer of $-2 \cdot 1 = -2$.

**Answer:** $-2$

## Problem 16 Solution

Let the angular measure of $\overset{\frown}{BD} = x$. Then the size of

$$\overset{\frown}{AC} = 360° - 50° - 50° - x = 260° - x,$$

and hence

$$\angle AEC = 50° = \frac{x-(260°-x)}{2}.$$

Solving for $x$ gives $x = 180°$.

**Answer:** 180

### Problem 17 Solution

Since the size of the regions are proportional let $x$ be the probability the spinner lands in the 1 region, so the other probabilities are $2x$, $3x$, $4x$, and $5x$ respectively.

Probabilities add up to 1, so

$$x + 2x + 3x + 4x + 5x = 15x = 1 \Rightarrow x = \frac{1}{15}.$$

Hence the probability it lands in the 1, 3, or 5 region is

$$\frac{1}{15} + \frac{3}{15} + \frac{5}{15} = \frac{9}{15} = \frac{3}{5}$$

so $M - N = 5 - 3 = 2$.

**Answer:** 2

### Problem 18 Solution

First note that the fourth polygon must have an interior angle of

$$360° - 90° - 90° - 60° = 120°$$

and hence must be a hexagon.

Therefore the perimeter of the entire shape is the sum of the perimeter of each of the four polygons minus the 4 shared edges. Note each shared edge is counted twice. Hence

$$3 \cdot (6 + 4 + 4 + 3 - 4 \cdot 2) = 27$$

is the perimeter.

**Answer:** 27

### Problem 19 Solution

Note the side view of the cup is a rectangle, and for the cup to be half full we have the diagram below (where the diagonal is parallel to the ground):

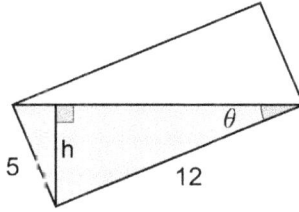

We then have that $\sin(\theta) = \dfrac{O}{H} = \dfrac{h}{12}$.

The diagonal has length $\sqrt{5^2 + 12^2} = 13$, so (using the larger right triangle) $\sin(\theta) = \dfrac{O}{H} = \dfrac{5}{13}$. Hence

$$\frac{h}{12} = \frac{5}{13} \Rightarrow h = 5 \cdot \frac{12}{13} \approx 4.6 < 5.$$

Therefore the smallest height possible is 5 cm.

**Answer:** 5

### Problem 20 Solution

We have $(\log_4 x)^2 + 2 = 3\log_4 x$ so with the substitution $y = \log_4 x$ we have

$$y^2 - 3y + 2 = (y-1)(y-2) = 0 \text{ so } y = 0 \text{ or } y = 1.$$

Therefore $\log_4 x = 1$ so $x = 4$ or $\log_4 x = 2$ so $x = 16$. Hence the sum of all the solutions is $4 + 16 = 20$.

**Answer:** 20

## 2.5 ZIML February 2018 Division H

Below are the solutions from the Division H ZIML Competition held in February 2018.
The problems from the contest are available on p.41.

### Problem 1 Solution
Using laws of logarithms we have

$$\log_2(x) + \log_2(x^2) = \log_2(x) + 2\log_2(x) = 3\log_2(x),$$

so dividing by 3 we need $\log_2(x) = 5$. Hence $x = 2^5 = 32$.

**Answer:** 32

### Problem 2 Solution
Since the order of the pairs does not matter, this is equivalent to each of the 6 women picking their partner. Picking one by one there are
$$6 \times 5 \times 4 \times 3 \times 2 \times 1 = 6! = 720$$
ways for them to do so, so there are 720 possible pairs.

**Answer:** 720

### Problem 3 Solution
The dimensions of the face with area 35 must be 7 and 5. As the face with area 42 shares a side with this face, its dimensions must be 7 and 6. =

Therefore, the dimensions of the prism are 5, 6, and 7. Hence

$$2 \times (5 \times 6 + 6 \times 7 + 7 \times 5) = 214$$

is the surface area of the prism.

**Answer:** 214

---

## Problem 4 Solution

The slope of $\overline{AB}$ must be the same as the slope of $\overline{AC}$. Thus

$$\frac{k-3}{3-1} = \frac{7-3}{k-1}.$$

Cross multiplying we have $(k-3)(k-1) = 8$ so $k^2 - 4k + 3 = 8$. Hence $k^2 - 4k - 5 = (k+1)(k-5) = 0$ so either $k = -1$ or $k = 5$. As $k > 0$, we must have $k = 5$.

**Answer:** 5

## Problem 5 Solution

Consider a right triangle $ABC$ with $\angle C = 90°$, $\angle A = \theta$, $AB = 19$, and $BC = 12$.

$$CA = \sqrt{19^2 - 12^2} = \sqrt{217}, \text{ and } \cos(\theta) = \frac{\sqrt{217}}{19}.$$

We have $14^2 = 184 < 217 < 225 = 15^2$ so $14 < \sqrt{217} < 15$. Checking $14.5^2 = 210.25$, we have $\sqrt{217} \approx 15$ so $P = 15$.

**Answer:** 15

## Problem 6 Solution

We first find the smallest positive number satisfying the requirements. Looking at a list of numbers with remainder 5 when divided by 16 we get

$$5, 21, 37, 53, 69, 85, 101, \ldots.$$

These numbers have remainders of

$$5, 0, 2, 4, 6, 1, 3, \ldots.$$

when divided by 7. (Notice a pattern of $+2$ every time since $5 + 2 = 7$ which has remainder 0.)

Hence we see that 85 is the smallest such number. As $\text{lcm}(7,16) = 7 \times 16 = 112$, the next smallest is $85 + 112 = 197$ which has 3-digits as needed.

**Answer:** 197

### Problem 7 Solution

Note that the number of different averages is the same as the number of different sums (as if you know the sum, you just divide by 5 to get the average).

The smallest possible sum is

$$1 + 2 + 3 + 4 + 5 = 15$$

while the largest possible sum is

$$16 + 17 + 18 + 19 + 20 = 90.$$

Every sum in between is possible, so there are $90 - 15 + 1 = 76$ possible sums and hence 76 possible averages.

**Answer:** 76

### Problem 8 Solution

Using the rational root theorem the integer roots are among $\pm 1, \pm 2$.

Checking we see $x = 1$ and $x = -2$ are roots, so $(x - 1)(x + 2) = x^2 + x - 2$ is a factor of the polynomial. Using polynomial long division we have

$$(x^4 - x^3 - 5x^2 + 3x + 2) \div (x^2 + x - 2) = x^2 - 2x - 1.$$

Hence by the quadratic formula (for quadratic $x^2 - 2x - 1$), $x = 1 \pm \sqrt{2}$ are the irrational roots, so $A + B = 1 + 2 = 3$.

**Answer:** 3

## Problem 9 Solution

It is straightforward using the Law of Sines. Solution without the Law of Sines is as follows.

The remaining angle is 105°. Draw the altitude from the vertex of 105°, and this altitude has length 6 based on a 30-60-90 triangle. Then the required side length is $6\sqrt{2} = \sqrt{72}$ based on a 45-45-90 triangle.

**Answer:** 72

## Problem 10 Solution

Consider the diagram below:

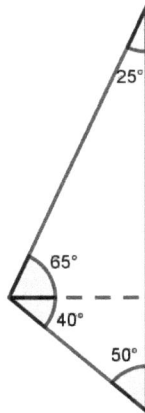

Label the triangle $\triangle ABC$ with $\angle A = 65° + 40° = 105°$, $\angle B = 25°$, $\angle C = 50°$, and altitude from vertex $A$ of length 5.

Let $D$ be the foot of the altitude from vertex $A$. Then

$$DB = \tan(65°) \times 5 \approx 2.14 \times 5 = 10.7$$
$$\text{and } CD = \tan(40°) \times 5 \approx 0.84 \times 5 = 4.2.$$

Thus, the height of the painting is approximately $10.7 + 4.2 = 14.9$ ft.

**Answer:** 14.9

## Problem 11 Solution

If $z = a + bi$ then $\bar{z} = a - bi$. Hence we have

$$(a + bi)^2 = (a - bi)^2$$
$$\Rightarrow a^2 - b^2 + 2abi = a^2 - b^2 - 2abi$$
$$\Rightarrow 4abi = 0.$$

Therefore $a = 0$ or $b = 0$.

Since $a^2 + b^2 \geq 1$, exactly one of $a, b$ is 0.

If $a = 0$ then $b = \pm 1, \pm 2, \ldots, \pm 10$, giving a total of 20 possible numbers.

Similarly if $b = 0$ we get another 20 possibilities.

There are thus $20 + 20 = 40$ total numbers meeting the requirements listed in the problem.

**Answer:** 40

## Problem 12 Solution

There are $36 + 2 = 38$ slots in total, with $4 + 4 = 8$ of the slots being black (they are $11, 13, 15, 17$ and $29, 31, 33, 35$).

Hence the probability of a slot that is odd and black is

$$\frac{8}{38} = \frac{4}{19}.$$

Therefore $Q - P = 19 - 4 = 15$.

**Answer:** 15

**Problem 13 Solution**

We know $AB = BC = CD = AD = 10$.

Let $AF = x$ so $DF = 10 - x$. Since $\overline{FE}$ and $\overline{FA}$ are both tangents, we have $FE = AF = x$. Similarly we have $CE = CB = 10$.

Therefore $\triangle CDF$ is a right triangle with sides 10, $10 - x$, and $10 + x$. Thus

$$(10 - x)^2 + 10^2 = (10 + x)^2$$
$$\text{or } x^2 - 20x + 100 + 100 = x^2 + 20x + 100.$$

hence $40x = 100$ and $x = 2.5$. Therefore $CF = 10 + 2.5 = 12.5$.

**Answer:** 12.5

**Problem 14 Solution**

$f(x) = \frac{1}{3}x - 3$ and swapping $x$ and $y$ we have

$$x = \frac{1}{3}y - 3 \Rightarrow y = 3(x + 3)$$

so $f^{-1}(x) = 3(x + 3)$.

If $f(x)f^{-1}(x) = 0$, then $f(x) = 0$ or $f^{-1}(x) = 0$. Therefore, $x = 9$ or $x = -3$ respectively.

Thus the sum of solutions is $9 + (-3) = 6$.

**Answer:** 6

**Problem 15 Solution**

Factoring, $120 = 2^3 \cdot 3 \cdot 5$. Hence $x = 2$, $x = 3$, or $x = 5$.

If $x = 2$, $y = 60 - 2 = 58$ which is not prime.

If $x = 3$, $y = 40 - 3 = 37$ which is prime.

If $x = 5$, $y = 24 - 5 = 19$ which is prime.

Hence the only pairs are $(3, 37)$ and $(5, 19)$, so the sum of all possible $y$ is $37 + 19 = 56$.

**Answer:** 56

## Problem 16 Solution
Note we can factor/rewrite the equation as

$$\frac{x^2 + 2xy + y^2}{x^3 y^2 + x^2 y^3 + x + y} = \frac{(x+y)^2}{((xy)^2 + 1)(x+y)} = \frac{x+y}{(xy)^2 + 1}.$$

Since $x + y = 4$ and $xy = 4 - 7 = -3$, we can then substitute to get that the expression is equal to $\dfrac{4}{9+1} = \dfrac{2}{5}$. Therefore $P + Q = 7$.

**Answer:** 7

## Problem 17 Solution
For $AC$ and $BC$ to have the same length, we need $C$ to lie on the perpendicular bisector of $\overline{AB}$. The midpoint of $\overline{AB}$ is

$$\left( \frac{1+4}{2}, \frac{3+2}{2} \right) = \left( \frac{5}{2}, \frac{5}{2} \right).$$

As the slope from $(1, 3)$ to $(4, 2)$ is $-\dfrac{1}{3}$ the perpendicular line has slope 3. Thus the equation of the line is

$$y - \frac{5}{2} = 3 \left( x - \frac{5}{2} \right) \text{ which simplifies to } y = 3x - 5.$$

Hence $m \times b = 3 \times (-5) = -15$.

**Answer:** $-15$

## Problem 18 Solution

We know that $2^{15}$ has $15 + 1 = 16$ factors.

Pairing the factors: $2^0$ with $2^{15}$, $2^1$ with $2^{14}$, etc., we see that each pair multiplies to $2^{15}$ and we have a total of $16 \div 2 = 8$ pairs.

Hence the product of all the factors is $(2^{15})^8 = 2^{120}$ so $M = 120$.

**Answer:** 120

## Problem 19 Solution

$\triangle ABC$ is an isosceles right triangle, so $\angle A = \angle B = 45°$. Hence each region of each circle inside the triangle is a sector with central angle $45°$. Thus these two regions can be combined to get a quarter circle.

For convenience suppose $AC = BC = 2$, so $AB = \sqrt{8} = 2\sqrt{2}$ and hence the circles have a radius of $\sqrt{2}$. Thus the quarter circle has area

$$\frac{1}{4} \times \pi \times (\sqrt{2})^2 = \frac{\pi}{2} \approx \frac{22}{7} \div 2 = \frac{11}{7}.$$

As the triangle has area $\dfrac{2 \times 2}{2} = 2$ the area inside the triangle but outside the circles is

$$\left(2 - \frac{11}{7}\right) \div 2 = \frac{3}{7} \div 2 = \frac{3}{14}$$

of the entire triangle. Hence $P + Q = 3 + 14 = 17$.

**Answer:** 17

## Problem 20 Solution

Substituting $y = \sqrt{x+2}$ we have $y^{-1} + y = \dfrac{10}{3}$.

Clearing denominators we have $3 + 3y^2 = 10y$ or

$$3y^2 - 10y + 3 = (3y - 1)(y - 3) = 0.$$

Hence $y = 3$ or $y = \dfrac{1}{3}$. If $\sqrt{x+2} = 3$ we have $x = 7$, which is not of the correct form. if $\sqrt{x+2} = \dfrac{1}{3}$ we have $x = -17/9$ so $P - Q = -17 - 9 = -26$.

**Answer:** $-26$

## 2.6   ZIML March 2018 Division H

Below are the solutions from the Division H ZIML Competition held in March 2018.

The problems from the contest are available on p.47.

### Problem 1 Solution

We have $y = x - 10$, so substituting we get $x^2 - 3(x - 10) = 28$ or

$$x^2 - 3x + 2 = (x - 2)(x - 1) = 0.$$

Hence $x = 2$ or $x = 1$. This gives the intersection points of $(2, -8)$ and $(1, -9)$. The sum of the $y$-coordinates of these points is $-8 - 9 = -17$.

**Answer:** $-17$

### Problem 2 Solution

Lines with the same slope are parallel, so the quadrilateral is a parallelogram. Label the points $ABCD$ and consider the base as $AB$ and the height $AE$ as in the diagram below.

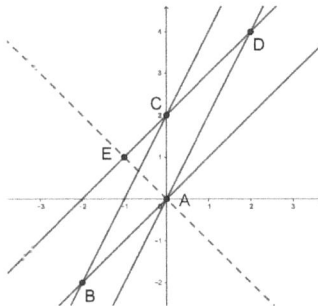

The point $A$ is $(0,0)$. $B$ is the intersection of $y = x$ and $y = 2x + 2$, which occurs when $2x + 2 = x$ or $x = -2$. Hence $B = (-2, -2)$. This gives $AB = \sqrt{2^2 + 2^2} = 2\sqrt{2}$.

Similarly, $E$ is the intersection of $y = x + 2$ with the line perpendicular to $y = x$, which is $y = -x$. Solving $x + 2 = -x$ we get $x = 1$ and thus $E = (-1, 1)$. Therefore $AE = \sqrt{1^2 + 1^2} = \sqrt{2}$.

Combining the information, the area of $ABCD$ is $2\sqrt{2} \cdot \sqrt{2} = 4$.

**Answer:** 4

## Problem 3 Solution

Let
$$f(x) = (5x - 3)^8 = ax^8 + bx^7 + \cdots + hx + i.$$
When $x = 1$ we have
$$f(1) = (5 \cdot 1 - 3)^8 = a + b + \cdots + h + i$$
so $a + b + \cdots + i = (5 - 3)^8 = 256$.

**Answer:** 256

## Problem 4 Solution

Consider the trapezoid as being made up of a $4 \times 2$ rectangle and a $4 \times 3$ triangle.

The rectangle will form a cylinder with radius 4 and height 2 when rotated. This cylinder has volume $\pi \cdot 4^2 \cdot 2 = 32\pi$.

The triangle will form a cone when rotated, with radius 4 and height 3. This cone has volume $\dfrac{1}{3} \cdot \pi \cdot 4^2 \cdot 3 = 16\pi$.

Thus the volume of the entire solid is $32\pi + 16\pi = 48\pi$.

**Answer:** 48

## Problem 5 Solution

In the worse case scenario, Bruce sells $10 to 40 customers and the remaining $662 - \$400 = \$262$ to the remaining customer which gives him a median sale of $10.

For the largest possible median (and hence bonus) we want the middle customer (the 21st) to be the largest sale possible. Therefore assume the first 20 customers were each sales of $10. This leaves $662 - $200 = $462 for the remaining 21 customers. Dividing this equally (so that the 21st is as large as possible), each of the remaining sales is $462 \div 21 = $22 for a median of $22.

This gives $22 - $10 = $12 difference from the largest and smallest possible bonuses.

**Answer:** 12

### Problem 6 Solution
Any number $\geq 2$ always has 1 and itself as a factor.

To have only one more factor, the third factor must be a prime. This implies that only numbers of the form $p^2$ for a prime $p$ have 3 factors.

Hence we are looking for the numbers $2^2$, $3^2$, $5^2$, $7^2$, $11^2$, $13^2$, and $17^2$ (the next would be $19^2 = 361$ which is too big). This is 7 numbers in total.

**Answer:** 7

### Problem 7 Solution
The square has area 36 so $EF = FG = 6$.

Further, as squares have all $90°$ angles, $\triangle ABC \sim \triangle FBG \sim AEF$ and all are 30-60-90 triangles.

This implies that $EF : AF = 1 : 2$ and $BF : FG = \sqrt{3} : 2$ so we have
$$AF = 2 \cdot EF = 2 \cdot 6 = 12$$
$$\text{and } BF = \frac{\sqrt{3}}{2} \cdot FG = \frac{\sqrt{3}}{2} \cdot 6 = 3\sqrt{3}.$$

Thus $AB = AF + BF = 12 + 3\sqrt{3}$ so $P + Q + R = 12 + 3 + 3 = 18$ is our answer.

**Answer:** 18

### Problem 8 Solution
Since there are no cubic or linear terms, $f(x) = ax^4 + bx^2 + c$ for some numbers $a, b, c$. Note then that

$$f(-5) = a(-5)^4 + b(-5)^2 + c$$
$$= a(5)^4 + b(5)^2 + c$$
$$= f(5)$$

so $f(-5) = 18$. (The graph has horizontal symmetry about the line $x = 0$.)

**Answer:** 18

### Problem 9 Solution
Using the triangle inequality we must have

$$AB + AC > BC, AB + BC > AC \text{ and } AC + BC > AB.$$

The third is automatically true, so we need

$$20 + 35 > \frac{z}{3}, 20 + \frac{z}{3} > 35 \Rightarrow 15 < \frac{z}{3} < 55 \Rightarrow 45 < z < 165.$$

However, we need $\frac{z}{3}$ to be an integer, so $z$ must be a multiple of 3.

Since $165 \div 3 = 55$ there are 54 multiples of 3 less than 165. $45 \div 3 = 15$ of them are less than or equal to 45, so in total there are $54 - 15 = 39$ possible values of $z$.

**Answer:** 39

## Problem 10 Solution

Combining the logarithms we have

$$\log_x((x-2)(3x+5)) = 2.$$

This is true if $(x-2)(3x+5) = x^2$. Expanding and combining like terms we have

$$2x^2 - x - 10 = (2x-5)(x+2) = 0.$$

Therefore $x = \dfrac{5}{2} = 2.5$ or $x = -2$. However, logarithms are not defined for negative values, so $x = 2.5$ is the only solution. Hence the sum of all solutions is just 2.5.

**Answer:** 2.5

## Problem 11 Solution

Let $P(1) = x$. As each successively larger triangle is made up of 4 smaller triangles we have

$$P(1) = P(2) = P(3) = P(8) = x,$$
$$P(4) = P(5) = P(9) = 4x,$$
$$\text{and } P(6) = P(7) = P(10) = 16x.$$

As all the probabilities sum to 1 we must have

$$1 = 4 \cdot (x) + 3 \cdot (4x) + 3 \cdot (16x) = 64x \Rightarrow x = \frac{1}{64},$$

so the probability we get an even number is

$$\frac{1}{64} + \frac{4}{64} + \frac{16}{64} + \frac{1}{64} + \frac{16}{64} = \frac{38}{64} = \frac{19}{32}$$

so $R + S = 19 + 32 = 51$.

**Answer:** 51

## Problem 12 Solution

Reaching the bottom of the tree takes 5 feet of the leash, so $10 - 5 = 5$ feet remain.

The tree has a diameter of 3, so circumference $\pi \times 3 = 3\pi \approx 9$.

Therefore the leash reaches $\approx \dfrac{5}{9}$ of the way around the tree. As $\dfrac{5}{9} = 0.\overline{5} = 55.\overline{5}\%$ we have $K = 56$.

**Answer:** 56

## Problem 13 Solution

For the number to be divisible by 18 it must be divisible by 2 and by 9.

First this implies $A$ is even. Further, for divisibility by 9 the sum of the digits,

$$2 + 0 + A + B + 1 + 8 + B + A = 2A + 2B + 11$$

needs to be a multiple of 9. Hence

$$2A + 2B + 11 = 2(A + B) + 11 = 18 \text{ or } 27 \text{ or } 36 \text{ or } 45 \text{ or } \cdots.$$

$2(A + B) + 11$ is odd with $2(A + B) + 11 \le 36 + 11 = 47$, so we must have $2(A + B) + 11$ is 27 or 45.

For the largest possible number, $2(A + B) + 11 = 45$ so $A + B = 17$. For $A$ to be even we have $A = 8, B = 9$. This gives the number 20891898.

**Answer:** 20891898

## Problem 14 Solution

We have that $2\sin(x)\cos(x) = \sin(2x)$ so our equation is

$$6\sin^2(2x) + \sin(2x) - 2 = 0.$$

Factoring we have

$$(2\sin(2x) - 1)(3\sin(3x) + 2) = 0 \Rightarrow \sin(2x) = \frac{1}{2} \text{ or } -\frac{2}{3}.$$

We want the smallest solution, so $\sin(2x) = \frac{1}{2}$ gives $2x = 30°$ and hence $x = D = 15°$ is the smallest such solution.

**Answer:** 15

**Problem 15 Solution**
Clearing denominators by multiplying by 35 we have

$$7x + 5y = 57.$$

Note that $5y$ will always have a ones digit of either 5 or 0, meaning we need $7x$ to have a ones digit of respectively 2 or 7.

This is only possible when $x = 1$ $(7 \cdot 1 = 7)$ or $x = 6$ $(7 \cdot 6 = 42)$. This leads to the solutions $(x, y) = (1, 10)$ or $(x, y) = (6, 3)$. $x^2 - y^2$ is largest for the second solution, with $x^2 - y^2 = 6^2 - 3^2 = 27$.

**Answer:** 27

**Problem 16 Solution**
Consider first $\angle A = \angle BAD$ and $\angle C = \angle BCD$.

$$\angle A = \frac{1}{2}\widehat{BCD} \text{ and } \angle C = \frac{1}{2}\widehat{BAD}$$

where, for example, $\widehat{BCD}$ is the arc containing $C$. Hence

$$\angle A + \angle C = \frac{1}{2}(\widehat{BCD} + \widehat{BAD}) = \frac{1}{2} \cdot 360° = 180°.$$

An identical argument gives $\angle B + \angle D = 180°$.

Thus we have $2x + 5y = 180°$ and $3x + 3y = 180°$. From the second equation we know $2x + 2y = 120°$ so subtracting this from the first equation we have $3y = 60°$ so $y = 20°$.

Finally, $3x + 3 \cdot 20° = 180°$ so $x = 40°$.

**Answer:** 40

### Problem 17 Solution
Since both the strings Dennis and May25 will appear in order, the password has length $5 + 6 = 11$ and is determined when we know which 6 places the letters in Dennis occupy.

As there are 11 positions in total, there are $\dbinom{11}{6} = \dfrac{11!}{6!5!} = 462$ total passwords.

**Answer:** 462

### Problem 18 Solution
Squaring the equation $\sqrt{x+3} = \sqrt{3x+2} - 1$ gives

$$x + 3 = 3x - 2 - 2\sqrt{3x-2} + 1 \text{ or } x - 2 = \sqrt{3x-2}.$$

Squaring one more time we have

$$x^2 - 7x + 6 = (x-6)(x-1) = 0,$$

and hence $x = 1$ or $x = 6$.

If $x = 1$ we have $\sqrt{1+3} - \sqrt{3-2} = 2 - 1 = 1 \neq -1$ so $x = 1$ is an extraneous solution.

Checking $x = 6$, $\sqrt{6+3} - \sqrt{3 \cdot 6 - 2} = 3 - 4 = -1$ so our only solution is $x = 6$.

**Answer:** 6

## Problem 19 Solution

We know $\sin^2 C + \cos^2 C = 1$ so

$$\left(\frac{4\sqrt{3}}{13}\right)^2 + (\cos C)^2 = 1 \Rightarrow (\cos C)^2 = 1 - \frac{48}{169} = \frac{121}{169}.$$

Hence $\cos C = \pm\frac{11}{13}$ so $\cos C = \frac{11}{13}$ as $\triangle ABC$ is acute. Now using the Law of Cosines we have

$$AB^2 = 24^2 + 25^2 - 2\cdot 24 \cdot 26 \cdot \frac{11}{13} = 1252 - 1056 = 196$$

so $AB = \sqrt{196} = 14$.

**Answer:** 14

## Problem 20 Solution

Rewriting the equation as a quadratic we have $z^2 + (-1+2i)z + (1+5i) = 0$. If $-1+i$ and $a+bi$ are solutions, the expression must factor:

$$z^2 + (-1+2i)z + (1+5i)$$
$$= (z+1-i)(z-a-bi)$$
$$= z^2 + (1-a+(-b-1)i)z + (a+b+(b-a)i).$$

Therefore

$$-1+2i = 1 - a + (-b-1)i \text{ and } 1+5i = a+b+(b-a)i.$$

Looking at the real and imaginary parts of the first equation we have $-1 = 1 - a$ so $a = 2$ and $2 = -b - 1$ so $b = -3$. Checking these also satisfy the second equation so our complex number is $2 - 3i$. For this number $a^2 + b^2 = 2^2 + (-3)^2 = 13$.

**Answer:** 13

## 2.7   ZIML April 2018 Division H

Below are the solutions from the Division H ZIML Competition held in April 2018.

The problems from the contest are available on p.55.

### Problem 1 Solution

Let $x, y, z$ be the number of cars Alice, Bob, Eve wash. We know $x + y + z = 600$, $3x = y + z$, and $2z = x + y$ from the given information.

Adding the first and third equations we get $x + y + 3z = 600 + x + y$ so canceling we have $3z = 600$ so $z = 200$

Putting this back into the second and third equations we have $3x = y + 200$ and $x + y = 400$.

The second says $x = 400 - y$ and thus after substituting we have $3(400 - y) = y + 200$ or $1200 - 3y = y + 200$.

Combining like terms gives $1000 = 4y$ or $y = 250$.

**Answer:** 250

### Problem 2 Solution

The sector is $\dfrac{270°}{360°} = \dfrac{3}{4}$ of a full circle, so the arc length of the outer edge is

$$\frac{3}{4} \cdot 2\pi \cdot 10 = 15\pi.$$

This will be the circumference of the base of the cone, hence the base of the cone has radius $15\pi \div (2\pi) = 7.5$.

Therefore the base of the cone has area $\pi \cdot (7.5)^2 = 56.25\pi$ and $K = 56.25$.

**Answer:** 56.25

## Problem 3 Solution

The graph of $y = x + C$ is a line with slope 1. The graph of $y = 4 - 2|x + 1|$ is

$$y = 4 + 2(x + 1) = 2x + 6 \text{ when } x < -1$$
$$\text{and } y = 4 - 2(x + 1) = -2x + 2 \text{ when } x > -1.$$

(It is a V-shape opening downwards.) Since the slope of the V-shape is steeper than the slope of the line, the line and absolute value graph intersect exactly once when the line hits the top of the V.

This point is occurs when $x = -1$ and

$$y = 2(-1) + 6 = -2(-1) + 2 = 4.$$

In this case $4 = -1 + C$ so $C = 5$.

**Answer:** 5

## Problem 4 Solution

There are $11! = 11 \cdot 10 \cdot 9 \cdots 1$ ways to rearrange the letters in PROBABILITY.

If we want the two B's and two I's to appear next to each other, we just need to arrange the 9 letters in PROBAILTY (as then we place the second B and I next to the other letters), which can be done in $9! \times 2! \times 2!$ ways, as there are 2 ways to arrange the B's and 2 ways to arrange the I's.

Therefore the probability is

$$\frac{9! \times 2 \times 2}{11!} = \frac{4}{11 \cdot 10} = \frac{2}{55}$$

and $Q - P = 55 - 2 = 53$.

**Answer:** 53

**Problem 5 Solution**

Calculating the prime factorizations we have

$$896 = 2^7 \cdot 7 \text{ and } 1875 = 3 \cdot 5^4 \Rightarrow \text{lcm}(896, 1875) = 2^7 \cdot 3 \cdot 5^4 \cdot 7$$

Thus $K$ is the smallest multiple of $2^7 \cdot 3 \cdot 5^4 \cdot 7$ that is a perfect cube.

To be a perfect cube, all the exponents must be a multiple of 3. Hence $K = 2^9 \cdot 3^3 \cdot 5^6 \cdot 7^3$.

Zeros at the end of $K$ come from powers of 10 and we write $K = 2^3 \cdot 3^3 \cdot 7^3 \cdot 10^6$ combining as many powers of 2 and 5 as we can.

Therefore $K$ ends in 6 zeros.

**Answer:** 6

**Problem 6 Solution**

First note that

$$\angle A = (\widehat{BC} - \widehat{DE}) \div 2 = (130° - 40°) \div 2 = 45°.$$

Therefore using the law of cosines we have

$$BC^2 = AB^2 + AC^2 - 2 \cdot AB \cdot AC \cdot \cos(\angle A)$$
$$= 10^2 + (8\sqrt{2})^2 - 2 \cdot 10 \cdot 8\sqrt{2} \cdot \frac{\sqrt{2}}{2}$$
$$= 100 + 128 - 160 = 68$$

and $BC^2 = 68$.

**Answer:** 68

## Problem 7 Solution

First we rewrite the equation as

$$\log_8(x) + \frac{1}{\log_8(x)} = \frac{10}{3}$$

so if $z = \log_8(x)$ we have the equation $z + \frac{1}{z} = \frac{10}{3}$. Clearing denominators we have

$$3z^2 + 3 = 10z \Rightarrow 3z^2 - 10z + 3 = (3z-1)(z-3) = 0$$

and hence $z = 3$ or $z = \frac{1}{3}$. Therefore

$$\log_8(x) = 3 \Rightarrow x = 8^3 = 512 \text{ or } \log_8(x) = \frac{1}{3} \Rightarrow x = \sqrt[3]{8} = 2.$$

Thus the difference between the largest and smallest solution is $512 - 2 = 510$.

**Answer:** 510

## Problem 8 Solution

Using the identity $\cos^2(\theta) - \sin^2(\theta) = \cos(2\theta)$ we can rewrite the equation as

$$2\cos^2(2\theta) + \cos(2\theta) = \cos(2\theta)(2\cos(2\theta) + 1) = 0.$$

Hence either $\cos(2\theta) = 0$ or $\cos(2\theta) = -\frac{1}{2}$. Since we want $0° \le \theta < 90°$, the first equation gives $2\theta = 90°$ or $\theta = 45°$ while the second gives $2\theta = 120°$ or $\theta = 60°$. Hence the sum of the solutions is $45° + 60° = 105°$ and our answer is 105.

**Answer:** 105

### Problem 9 Solution

The cube has volume $4^3 = 64$ cm$^3$.

The square pyramid has a volume of $\frac{1}{3} \cdot 4^2 \cdot 2 = \frac{32}{3} \approx 10.66$ cm$^3$.

Therefore the total volume of the new solid is $74.66 \approx 75$ cm$^3$.

**Answer:** 75

### Problem 10 Solution

Notice $18x - 45 = 9(2x - 5)$ so we can rewrite the equation as $(2x - 5)^3 = 9(2x - 5)$.

$2x - 5 = 0$ leads to $x = 2.5$ as one solution (which is not an integer). Else we can divide by $(2x - 5)$ on both sides to get

$$(2x - 5)^2 = 9 \Rightarrow 2x - 5 = \pm 3 \Rightarrow x = \frac{5 \pm 3}{2} \Rightarrow x = 4 \text{ or } x = 1.$$

Thus the sum of the integer solutions is $1 + 4 = 5$.

**Answer:** 5

### Problem 11 Solution

If we want the last two digits of $91^{91}$ we need to find a pattern of the last two digits.

We have $91^2 = 8281$ which has last two digits 81. Then $91 \cdot 81 = 7371$ with last two digits 71. Continuing we see the pattern is

$$91, 81, 71, 61, 51, 41, 31, 21, 11, 01$$

which then starts repeating.

As this pattern has a length of 10, and $91 = 9 \cdot 10 + 1$, the last two digits of $91^{91}$ are the first term of the pattern: 91.

**Answer:** 91

## Problem 12 Solution

The four circles have radii of 1, 2, 3, and 4, so their respective areas are $\pi$, $4\pi$, $9\pi$, and $16\pi$. Thus, the probability of landing in each of the 4 regions is

$$5 \text{ Point:} \quad \frac{\pi}{16\pi} = \frac{1}{16}$$
$$3 \text{ Point:} \quad \frac{4\pi - \pi}{16\pi} = \frac{3}{16}$$
$$2 \text{ Point:} \quad \frac{9\pi - 4\pi}{16\pi} = \frac{5}{16}$$
$$1 \text{ Point:} \quad \frac{16\pi - 9\pi}{16\pi} = \frac{7}{16}.$$

For the two shots to have a combined total of 6 points, they were both either 3 point shots or one was 5 points and the other 1 point. Hence the probability is

$$\frac{3}{16} \cdot \frac{3}{16} + \frac{1}{16} \cdot \frac{7}{16} + \frac{7}{16} \cdot \frac{1}{16} = \frac{9 + 7 + 7}{256} = \frac{23}{256}.$$

As $\dfrac{23}{256} \approx 0.08984 = 8.98\%$ we have $S = 9$.

**Answer:** 9

## Problem 13 Solution

Call the entire triangle $ADE$ (with $B$ on $\overline{DE}$ and $C$ on $\overline{AE}$). $\triangle ABC$ and $\triangle ABE$ have the same height and also $\triangle ABE$ and $\triangle ADE$ share the same height, so

$$\frac{[ABC]}{[ABE]} = \frac{3}{4} \text{ and } \frac{[ABE]}{[ADE]} = \frac{3}{4}.$$

(Here, for example, $[ABC]$ denotes the area of $\triangle ABC$.) Therefore

$$\frac{[ABC]}{[ADE]} = \frac{[ABC]}{[ABE]} \cdot \frac{[ABE]}{[ADE]} = \frac{3}{4} \cdot \frac{3}{4} = \frac{9}{16}.$$

Since $\triangle ADE$ has area $16 \cdot 2 = 32$ (it is made up of 16 small equilateral triangles), $\triangle ABC$ has area 18.

**Answer:** 18

### Problem 14 Solution

Trying to find a pattern we have

$$(\sqrt{3}+i)^2 = 2 + 2i\sqrt{3} = 2(1 + i\sqrt{3})$$

and thus

$$(\sqrt{3}+i)^3 = 2(1+i\sqrt{3})(\sqrt{3}+i) = 2(\sqrt{3}+i+3i+i^2\sqrt{3}) = 2^3 i.$$

Hence $(\sqrt{3}+i)^6 = (2^3 i)^2 = -2^6$. This gives

$$(\sqrt{3}+i)^{81} = (\sqrt{3}+i)^{78+3}$$
$$= \left((\sqrt{(3)}+i)^6\right)^{13} (\sqrt{3}+i)^3$$
$$= \left(-2^6\right)^{13} \cdot 2^3 i = -2^{81} i.$$

Therefore $K + A + B = 81 + 0 - 1 = 80$.

**Answer:** 80

### Problem 15 Solution

First complete the square to get

$$x^2 - 6x + y^2 - 20y + 93 = (x-3)^2 + (y-10)^2 - 16$$

so the circle has equation $(x-3)^2 + (y-10)^2 = 4^2$ with center $(3, 10)$ and radius 4. Notice that the center of the circle $(3, 10)$ is on the line $y = 2x + 4$.

This means that the line segment inside the circle is a diameter, hence cuts the circle in half. Therefore the region has area $\pi 4^2 \div 2 = 8\pi$ and $S = 8$.

**Answer:** 8

---

## Problem 16 Solution

We use the Polynomial Remainder Theorem. This tell us that $P(-1) = -10$ and $P(-2) = -37$. Hence

$$2(-1)^3 - a(-1)^2 - b(-1) - 1 = -10 \Rightarrow b - a = -7$$

and similarly

$$2(-2)^3 - a(-2)^2 - b(-2) - 1 = -37 \Rightarrow 2b - 4a = -20.$$

Doubling the first equation gives $2b - 2a = -14$ and subtracting this from the second equation we have $-2a = -6$ so $a = 3$. Thus $b - 3 = -7$ and $b = -4$.

Therefore $P(x) = 2x^3 - 3x^2 + 4x - 1$ and $P(2) = 2 \cdot 2^3 - 3 \cdot 2^2 + 4 \cdot 2 - 1 = 16 - 12 + 8 - 1 = 11$.

**Answer:** 11

## Problem 17 Solution

Let $x$ denote the side length of the square (so all 3 squares have side length $x$). Then the other leg of each 30-60-90 triangle is $x\sqrt{3}$.

Pairing up the triangles, the area of the octagon is the area of the 3 squares plus the area of 4 of the rectangles. Thus

$$3x^2 + 4(x^2\sqrt{3}) = 6 + 8\sqrt{3} \Rightarrow x^2 = \frac{6 + 8\sqrt{3}}{3 + 4\sqrt{3}} = 2$$

so the side length of the squares is $\sqrt{2}$. As $\sqrt{2} \approx 1.414$, rounded to the nearest tenth this is 1.4.

**Answer:** 1.4

## Problem 18 Solution

$\triangle ABC \sim \triangle DEF$ as they have the same angles. Therefore their sides are proportional and

$$\frac{30}{x+2} = \frac{15}{x-2} \Rightarrow 30(x-2) = 15(x+2).$$

Simplifying we have $2x - 4 = x + 2$ and $x = 6$.

**Answer:** 6

## Problem 19 Solution

First ignore the numbers on each ball. In the 5 picks, Izzy chooses 3 yellow balls. Hence there are

$$\binom{5}{3} = \frac{5!}{3! \cdot 2!} = 10$$

ways to decide which 3 of the 5 balls chosen are yellow (the remaining two are purple).

For each of the yellow balls chosen, it could either be numbered 1 or 2, so there are $2 \cdot 2 \cdot 2 = 2^3 = 8$ choices for the numbering of the yellow balls.

Similarly there are $3 \cdot 3 = 9$ choices for the numbering of the purple balls.

In total this gives $10 \cdot 8 \cdot 9 = 720$ outcomes.

**Answer:** 720

## Problem 20 Solution

Rationalizing the denominator we have

$$T = \frac{6}{\sqrt{7}-1} \cdot \frac{\sqrt{7}+1}{\sqrt{7}+1} = \sqrt{7}+1.$$

As $\sqrt{7} \approx 2.64$, $\lfloor T \rfloor = 3$ and $\{T\} = \sqrt{7} - 2$. Hence

$$\lfloor T \rfloor + (2 + \sqrt{7})\{T\} = 3 + (\sqrt{7} + 2)(\sqrt{7} - 2) = 3 + 3 = 6$$

as our final answer.

**Answer:** 6

## 2.8 ZIML May 2018 Division H

Below are the solutions from the Division H ZIML Competition held in May 2018.
The problems from the contest are available on p.63.

### Problem 1 Solution
We are given that $RS = PQ = 8$. Since the area of the parallelogram is 48, $ST = 48 \div 8 = 6$.

As $T$ is the midpoint, $PT = 8 \div 2 = 4$, so using the Pythagorean theorem we have that

$$PS = RQ = \sqrt{4^2 + 6^2} = \sqrt{52} = 2\sqrt{13}.$$

The perimeter is

$$2(8 + 2\sqrt{13} = 16 + 4\sqrt{13}$$

and $A + B + C = 16 + 4 + 13 = 33$.

**Answer:** 33

### Problem 2 Solution
If $\log_5(x^2 - 4) = 5$ we have $x^2 - 4 = 5^5 = 3125$.

Hence $x = \sqrt{3129}$. $50^2 = 2500$ and $60^2 = 3600$ so $\sqrt{3129}$ is between 50 and 60. Checking, $55^2 = 3025$ and $56^2 = 3136$.

Therefore $\sqrt{3129} \approx 56$ rounded to the nearest integer.

**Answer:** 56

### Problem 3 Solution
We know at least $27 \div 3 = 9$ of the collector's edition cars are in their unopened boxes. Thus, at most $23 - 9 = 14$ regular toy cars

---

are in their unopened boxes and so at least $33 - 14 = 19$ regular toy cars do not have their boxes anymore.

**Answer:** 19

## Problem 4 Solution
The equation is equivalent to

$$\sqrt{3x+2} = 3 - \sqrt{3x-2}.$$

Squaring both sides we obtain

$$3x+2 = 3x+7-6\sqrt{3x-2} \Rightarrow 6\sqrt{3x-2} = 5.$$

Squaring once more we have $36(3x-2) = 25$, which has solution $x = \dfrac{97}{108}$.

This is not an extraneous solution, so the sum of all solutions to the equation is $\dfrac{97}{108} \approx 0.9$.

**Answer:** 0.9

## Problem 5 Solution
Let $O$ be the center of the sphere and $I$ be the center of one of the bases of the cylinder.

Consider $\triangle AOI$, where $A$ is some point on the circumference of the circle with center $I$. $\triangle AOI$ is a right triangle with hypotenuse 26, and legs $20 \div 2 = 10$ and $r$, where $r$ is the radius of the base of the cylinder.

Since $(10, 24, 26)$ is a Pythagorean triple, $r = 24$.

**Answer:** 24

### Problem 6 Solution

For the number to have 15 factors, it must be of the form $p^{14}$ for a prime $p$ or $p^4 q^2$ for primes $p, q$ (as $15 = 15 = 5 \cdot 3$).

However, if it is a multiple of 22, it contains 2 and 11 as factors, so it must be of the form $p^4 q^2$.

Any number of this form is a perfect square, so both numbers $2^4 \cdot 11^2$ and $2^2 \cdot 11^4$ match the descriptions given by Barry, Carrie, and Mary. These have difference

$$2^2 \cdot 11^4 - 2^4 \cdot 11^2 = 2^2 \cdot 11^2 \cdot (11^2 - 2^2) = 484 \cdot 117 = 56628$$

which is our final answer.

**Answer:** 56628

### Problem 7 Solution

We can factor

$$x^3 + 4x^2 + x + 4 = x^2(x+4) + x + 4 = (x^2 + 1)(x+4)$$

and similarly

$$x^3 - 8 = (x - 2)(x^2 + 2x + 4).$$

$x^2 + 1 > 0$ for all $x$, and the discriminant of $x^2 + 2x + 4$ is $2^2 - 4(1)(4) = -12 < 0$.

Therefore the equation has only 2 real solutions. (The solutions are $x = 1$ and $x = 2$).

**Answer:** 2

### Problem 8 Solution

We want $f(n+1) + f(n) = n$ or

$$(n+1)^2 - 1 + n^2 - 1 = n.$$

Expanding we have $2n^2 + 2n - 1 = n$ or $2n^2 + n - 1 = 0$. This factors as $(2n - 1)(n + 1) = 0$ so $n = 0.5$ or $n = -1$.

The only integer is $n = -1$.

**Answer:** $-1$

## Problem 9 Solution

From the diagram we have

$$\sin(\theta) = \frac{6}{10} = \frac{3}{5} \text{ and } \cos(\theta) = \frac{8}{10} = \frac{4}{5}.$$

Thus

$$\frac{1}{\sin(\theta)} - \frac{1}{\cos(\theta)} = \frac{5}{3} - \frac{5}{4} \approx 1.66 - 1.25 \approx 0.41$$

Therefore rounded to the nearest tenth the expression is approximately 0.4.

**Answer:** 0.4

## Problem 10 Solution

Using polynomial long division note that

$$\frac{x^3 - 2x^2 - 5x + 6}{x + 2} = x^2 - 4x + 3.$$

Hence for every value of $x \neq -2$, the given function is the same as the function $y = x^2 - 4x + 3$, a quadratic.

This quadratic takes a minimum value at its vertex. The $x$-value of the vertex is $-\dfrac{-4}{2 \cdot 1} = 2$. For this $x$, $y = 2^2 - 4 \cdot 2 + 3 = -1$. Hence $-1$ is the minimum value.

**Answer:** $-1$

## Problem 11 Solution

As the order does not matter, there are $\binom{10}{3} = \dfrac{10!}{3! \cdot 7!} = 120$ ways for him to buy the chocolate and $\binom{5}{2} = \dfrac{5!}{2! \cdot 3!} = 10$ ways for him to buy the gum.

In total this is $120 \cdot 10 = 1200$ different collections of candy.

**Answer:** 1200

## Problem 12 Solution

Substituting $y = \dfrac{x\sqrt{3}}{3} + \dfrac{4\sqrt{3}}{3}$ into the equation of the circle we have

$$x^2 + \left(\frac{x\sqrt{3}}{3} + \frac{4\sqrt{3}}{3}\right)^2 = 16 \Rightarrow x^2 + \frac{x^2}{3} + \frac{8x}{3} + \frac{16}{3} = 16.$$

Clearing denominators we have $x^2 + 3x^2 + 8x + 16 = 48$ which simplifies to $x^2 + 2x - 8 = 0$.

Factoring we have $(x+4)(x-2) = 0$ so $x = -4$ and $x = 2$ are solutions. This gives points $(-4, 0)$ and $(2, 2\sqrt{3})$ as the intersection points.

If we look at the (right) triangle formed by $(0,0)$, $(2,0)$, and $(2, 2\sqrt{3})$ we recognize this as a 30-60-90 triangle. Thus the shaded sector has a central angle of $180° + 60° = 240°$. This implies the sector has area

$$\frac{240°}{360°} \cdot \pi \cdot 4^2 = \frac{2}{3} \cdot 16\pi = \frac{32}{3}\pi.$$

Therefore $P + Q = 32 + 3 = 35$.

**Answer:** 35

---

## Problem 13 Solution

Using the Pythagorean identity, we have

$$6\left(1 - \sin^2(x)\right) = 4 - \sin(x).$$

Let $u = \sin(x)$, then the equation becomes $6\left(1 - u^2\right) = 4 - u$, so $6u^2 - u - 2 = 0$.

Factoring, $(2u + 1)(3u - 2) = 0$, so $u = \dfrac{2}{3}, -\dfrac{1}{2}$. Since $\sin(x) > 0$ for $90° < x < 180°$, we have $\sin(x) = \dfrac{2}{3} \approx 0.67$.

**Answer:** 0.67

## Problem 14 Solution

Factoring we have

$$5! + 6! + 7! = 5!(1 + 6 + 6 \cdot 7) = 5! \cdot 49 = 5! \cdot 7^2.$$

Similarly

$$7! + 8! + 9! = 7!(1 + 8 + 72) = 7! \cdot 9^2.$$

As $7! = 7 \cdot 6 \cdot 5!$, the greatest common factor is $5! \cdot 7 = 120 \cdot 7 = 840$.

**Answer:** 840

## Problem 15 Solution

Solving for $x$ we have $x = 100 - 3y$.

Since $x > 0$ we need $100 - 3y > 0$ or $3y < 100$ or $y < 33.\overline{3}$.

As $y > 0$ with $y$ an integer, we see $y = 1, 2, \ldots, 33$ all work.

Each of these gives a unique point $(x, y)$ (for example, $(97, 1)$ or $(94, 2)$) so there are 33 lattice points in total.

**Answer:** 33

## Problem 16 Solution

Let the angles have measures $3x, 4x, 5x$ so $3x + 4x + 5x = 180°$ and hence $x = 180° \div 12 = 15°$.

Hence the three angles are $45°, 60°, 75°$. The smallest side is across from the smallest angle $45°$ and similarly the middle side is across from the $60°$ angle.

If $m$ denotes the length of the middle side, using the Law of Sines we have

$$\frac{5}{\sin(45°)} = \frac{m}{\sin(60°)} \Rightarrow m = \frac{5 \cdot \sin(60°)}{\sin(45°)}.$$

Using $\sin(45°) = \frac{\sqrt{2}}{2}, \sin(60°) = \frac{\sqrt{3}}{2}$ we have

$$m = \frac{5 \cdot \sqrt{3}}{\sqrt{2}} = \frac{5\sqrt{6}}{2}.$$

Therefore $P + Q + R = 5 + 6 + 2 = 13$.

**Answer:** 13

## Problem 17 Solution

There are $2^5 = 32$ total outcomes. Consider cases based on how many heads Jason gets.

If he gets 5 heads, the one outcomes $HHHHH$ contains a $HHH$ as needed.

If he gets 4 heads, the outcomes $HHHHT, HHHTH, HTHHH$, and $THHHH$ all contain $H$ (for 4 heads, only $HHTHH$ does not work).

Lastly, if he gets 3 heads, the outcomes $HHHTT, THHHT$, and $TTHHH$ are the only ones that work (as the three heads must be consecutive). Therefore $1+4+3 = 8$ of the outcomes contain $HHH$.

This has probability $\dfrac{8}{32} = \dfrac{1}{4}$ so $P+Q = 1+4 = 5$.

**Answer:** 5

### Problem 18 Solution

The graph of $y = |x| - 6$ is given by $y = x - 6$ when $x \geq 0$ and $y = -x - 6$ when $x < 0$.

Similarly the graph of $y = -|x - 2| + 2$ is given by $y = -x + 4$ when $x \geq 2$ and $y = x$ when $x < 2$.

From these equations we have

$$x - 6 = -x + 4 \Rightarrow x = 5 \text{ and } -x - 6 = x \Rightarrow x = -3$$

are the $x$-values of the two intersection points. Plotting these points $(5, -1)$ and $(-3, 3)$ with the vertices $(0, -6)$ and $(2, 2)$ of the absolute value graphs gives the rectangle shown below:

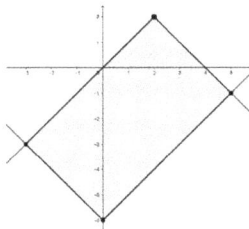

The distance formula gives us dimensions of

$$\sqrt{(0 - (-3))^2 + (-6 - (-3))^2} = \sqrt{18}$$

and

$$\sqrt{(5 - 0)^2 + (-1 - (-6))^2} = \sqrt{50}.$$

This gives

$$\sqrt{18} \cdot \sqrt{50} = 3\sqrt{2} \cdot 5\sqrt{2} = 15 \cdot 2 = 30.$$

as the area of the rectangle.

**Answer:** 30

## Problem 19 Solution

Since $z = 3 - 2i$ is one root, $\bar{z} = 3 + 2i$ is also a root. Thus

$$(z - (3 - 2i))(z - (3 + 2i)) = z^2 - 6z + 13$$

is a factor of $p(z)$. Using long division, we have

$$p(z) = (z^2 - 6z + 13)(z^2 - 6z + 8).$$

As $(z^2 - 6z + 8) = (z - 2)(z + 4)$, $z = 2, 4$ are also roots of $p(z)$.

Therefore, the sum of all real solutions is $2 + 4 = 6$.

**Answer:** 6

## Problem 20 Solution

Without loss of generality, assume $AB = 1$. Then the diagonals of each of the faces of the cube have length $\sqrt{2}$ and thus $AD = \sqrt{3}$.

$ABE$ is an isosceles triangle with $AE = BE = \dfrac{\sqrt{3}}{2}$, $AB = 1$. Using the Law of Cosines, we have

$$AB^2 = AE^2 + BE^2 - 2 \cdot AE \cdot BE \cdot \cos(\angle AEB),$$

so $1 = \dfrac{3}{2} - \dfrac{3}{2}\cos(\angle AEB)$ and $\cos(\angle AEB) = \dfrac{1}{3} \approx 0.33.$

**Answer:** 0.33

## 2.9  ZIML June 2018 Division H

Below are the solutions from the Division H ZIML Competition held in June 2018.
The problems from the contest are available on p.71.

### Problem 1 Solution
Suppose the radius of the cylinder is 1. Thus the diameter is 2 and hence the height is 4. Further, the radius of the half-sphere is 3.

For the volume, the half-sphere has volume $\frac{2}{3}\pi \cdot 3^3 = 18\pi$ and the cylinder has volume $\pi 1^2 \cdot 4 = 4\pi$. Therefore the total volume is $18\pi + 4\pi = 22\pi$.

For the surface area, half of a sphere has surface area $2\pi 3^2 = 18\pi$. The lateral surface area of the cylinder is $2\pi \cdot 1 \cdot 4 = 8\pi$. Lastly we have the area of one circle with radius 3, which is $\pi \cdot 3^2 = 9\pi$. Therefore the total surface area is $18\pi + 8\pi + 9\pi = 35\pi$.

Thus the ratio of volume to surface area is $22\pi : 35\pi = 22 : 35$ so $P + Q = 22 + 35 = 57$.

**Answer:** 57

### Problem 2 Solution
August can choose which books to give to Bernard in $\binom{6}{2} = 15$ different ways. Then Bernard can choose which books to give to August in $\binom{4+2}{2} = 15$ different ways.

Thus, they could have exchanged books in $15 \times 15 = 225$ different ways.

**Answer:** 225

## Problem 3 Solution

We can rewrite $f(x)$ as

$$f(x) = x^3 - 9x^2 + 27x - 27 + 4 = (x-3)^3 + 4,$$

so $f^{-1}(x) = \sqrt[3]{x-4} + 3$. Thus

$$f(23) - f^{-1}(-23) = \left((23-3)^3 + 4\right) - \left(\sqrt[3]{-23-4} + 3\right)$$
$$= 8004 - 0 = 8004.$$

**Answer:** 8004

## Problem 4 Solution

Since the number of bacteria in a sample doubles every 20 minutes, after 1 hour he would have $2 \times 2 \times 2 = 8$ times the bacteria. Hence, after $t$ hours Charley has $50 \times 200 \times 8^t$ bacteria.

We need to find $t$ such that $50 \times 200 \times 8^t \geq 5 \times 10^7$. This is equivalent to

$$8^t \geq \frac{5 \times 10^7}{10^4} = 5 \times 10^3 = 5000.$$

Note $8^4 = 4096$, so he needs to wait for at least 5 hours to get the bacteria he needs.

**Answer:** 5

## Problem 5 Solution

During the first leg of the trip the boat travels $25 \times 2 = 50$ miles, and during the second leg of the trip the boat travels for $20 \times 6 = 120$ miles. The diagram below shows the trajectory of the boat assuming it started at the origin.

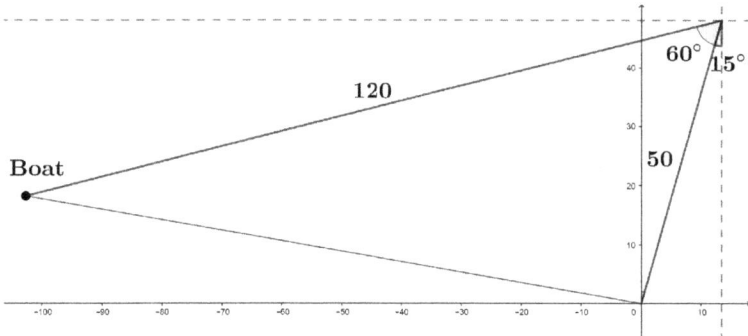

Note the two legs the boat traveled so far form an angle of $75° - 15° = 60°$ between them. Using the law of cosines we have the distance, $m$, from the point of origin to the boat is such that

$$m^2 = 50^2 + 120^2 - 2 \cdot 50 \cdot 120 \cdot \cos(60°)$$

$$= 16900 - 12000 \cdot \frac{1}{2}$$

$$= 10900$$

Thus, the distance from the origin to the boat is $\sqrt{10900}$ miles and $m = 10900$.

**Answer:** 10900

## Problem 6 Solution

Using the change of base formula we can turn the equation into

$$\log_3(x) + \frac{\log_3(x^2)}{\log_3(9)} + \frac{\log_3(x^3)}{\log_3(27)} = 9,$$

hence

$$\log_3(x) + \frac{\log_3(x^2)}{2} + \frac{\log_3(x^3)}{3} = 9$$

$$\Rightarrow 6\log_3(x) + 3\log_3(x^2) + 2\log_3(x^3) = 54,$$

so combining like terms we have $\log_3(x^6) = 18$. Thus $x^6 = 3^{18}$, and $x = \pm 3^3 = \pm 27$. Since $\log_3(x)$ is only defined when $x > 0$, the only real solution to the equation is $x = 27$.

**Answer:** 27

### Problem 7 Solution
Since $ABCDE$ is a pentagon, the sum of its internal angles is equal to $180° \times (5 - 2) = 540°$. This means

$$\angle EAC = \angle ABC = (540° - 110° - 102° - 170°) \div 2 = 79°.$$

The exterior angle at vertex $A$ is supplementary to $\angle EAB$, so its measure is $180° - 79° = 101°$.

**Answer:** 101

### Problem 8 Solution
We can factor $18000 = 2^4 \times 3^2 \times 5^3$ and $21600 = 2^5 \times 3^3 \times 5^2$. Thus, $\gcd(18000, 21600) = 2^4 \times 3^2 \times 5^2$.

Hence any perfect square that divides both numbers is of the form $2^a \times 3^b \times 5^c$, with $a = 0, 2, 4$, $b = 0, 2$ and $c = 0, 2$.

Therefore, there are $3 \times 2 \times 2 = 12$ perfect squares that divide both 18000 and 21600.

**Answer:** 12

### Problem 9 Solution
There are 13 ways of getting all cards with the same rank,

$$12 \times \binom{4}{3} \times \binom{4}{1} = 192$$

ways of getting three cards of the same rank and one $A$,

$$12 \times \binom{4}{2} \times \binom{4}{2} = 432$$

ways of getting two cards of the same rank and 2 $A$ cards, and

$$12 \times \binom{4}{1} \times \binom{4}{3} = 192$$

ways of getting 3 $A$ cards and one other card of a different rank.

Thus, there are $13 + 192 + 432 + 192 = 829$ possible winning hands (out of $\binom{52}{4} = 270725$ possible hands).

**Answer:** 829

### Problem 10 Solution

The maximum area will be attained when the center of the third circle is the same as the center of the first circle. Then the overlap of the three circles is just the overlap of the first two circles. This overlap is made up of two equilateral triangles of side 1 and four equal segments (a segment is the region between a cord and an arc of a circle).

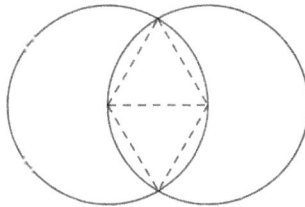

Each equilateral triangle has area $\dfrac{\sqrt{3}}{4}$, and each segment has area $\dfrac{\pi}{6} - \dfrac{\sqrt{3}}{4}$. Thus, the whole overlap has area

$$2\left(\frac{\sqrt{3}}{4}\right) + 4\left(\frac{\pi}{6} - \frac{\sqrt{3}}{4}\right) = \frac{2\pi}{3} - \frac{\sqrt{3}}{2} = \frac{4\pi - 3\sqrt{3}}{6},$$

so $A + B + C = 4 - 3 + 6 = 7$.

**Answer:** 7

## Problem 11 Solution

$f(x)$ is a parabola that opens up. The $x$-coordinate of its vertex is
$$-\frac{-36}{2 \cdot 3)} = 6.$$

Thus if $m \geq 6$, $f(x)$ has a well defined inverse. Hence the smallest value of $m$ is 6.

**Answer:** 6

## Problem 12 Solution

Looking at remainders $\pmod 7$ the sequence is

$$3, 5, 5, 3, 4, 2, 2, 4, 3, 5, 5, 3, 4, 2, 2, \ldots.$$

Note the sequence follows a pattern that repeats every 8 terms.

Since $2018 \div 8 = 252$ with remainder 2, the remainder of $G_{2018}$ when dividing by 7 is the $2^{\text{nd}}$ number in the repeating pattern, that is, 5.

**Answer:** 5

## Problem 13 Solution

We can rearrange the shaded region like in the diagram below.

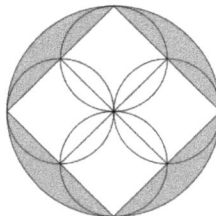

Thus, the shaded area is equal to the area of the circle minus a square with diagonal 8. That is, the shaded area is equal to $16\pi - 8^2 \div 2 = 16\pi - 32$. Hence, $A + B = -16$.

**Answer:** $-16$

### Problem 14 Solution
Using double angle trigonometric identities, we can rewrite the left side of the equation as

$$\cos(x) - \sin(x)\sin(2x) - \cos(2x)$$
$$= \cos(x) - \sin(x)(2\sin(x)\cos(x)) - (2\cos^2(x) - 1)$$
$$= \cos(x) - 2\sin^2(x)\cos(x) - 2\cos^2(x) + 1$$
$$= \cos(x) - 2(1 - \cos^2(x))\cos(x) - 2\cos^2(x) + 1$$
$$= \cos(x) - 2\cos(x) + 2\cos^3(x) - 2\cos^2(x) + 1$$
$$= 2\cos^3(x) - 2\cos^2(x) - \cos(x) + 1$$

Which can be factored as $(\cos(x) - 1)(2\cos^2(x) - 1)$, giving $\cos(x) = 1, \pm\dfrac{\sqrt{2}}{2}$, and $x = 0°, 45°, 135°$.

Thus, the difference between the largest and smallest solutions is $135 - 0 = 135$.

**Answer:** 135

### Problem 15 Solution
Let $n$ be the smallest of the 7 numbers. Then the sum of all 7 numbers is

$$n + (n+1) + \cdots + (n+6) = 7n + 21 = 7(n+3).$$

Note $154 = 7 \times 22$, so we are looking for the smallest 3-digit integer $n$ such that 22 divides $n + 3$. The smallest 3-digit multiple of 22 is 110, so $n = 107$.

**Answer:** 107

---

## Problem 16 Solution

If $x \geq -4$, the equation turns into

$$|2x - (x+4)| = 16,$$

so $|x - 4| = 16$ and $x = 20$ ($x = -12$ is extraneous).

If $x \leq -4$, the equation turns into

$$|2x + (x+4)| = 16,$$

so $|3x + 4| = 16$ and $x = -\dfrac{20}{3}$ ($x = 4$ is extraneous).

Thus, the sum of all real solutions to the equation is

$$20 - \frac{20}{3} = \frac{40}{3} \approx 13.33.$$

Rounded to the nearest tenth this is 13.3.

**Answer:** 13.3

## Problem 17 Solution

There are 3 kinds of parallelograms that can be formed, depending on which 2 sides of the big triangle we choose to make the parallel lines, as shown on the diagram below.

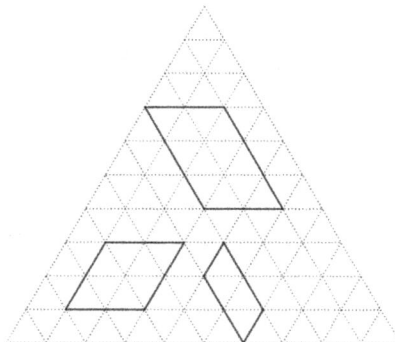

Once the direction of the parallelogram is chosen, we just need to choose 2 pairs of different parallel lines to those on the grid to determine a unique parallelogram. This can be achieved in $\binom{10}{2} = 45$ ways.

Therefore, there are $3 \times 45 \times 45 = 6075$ different parallelograms.

**Answer:** 6075

### Problem 18 Solution

Let $S$ denote the sum. Then $2S$ is

$$\sum_{k=1}^{10} (-1)^k 2^k = \sum_{k=1}^{10} (-2)^k = \frac{(-2)^{11} - 1}{-2 - 1} - 1 = \frac{2049}{3} - 1 = 682$$

using the formula for a geometric series. Thus, $S = 682 \div 2 = 341$.

**Answer:** 341

### Problem 19 Solution

As $15° = 45° - 30°$, we can use the values of sine and cosine of $30°$ and $45°$ to find the $\sin(15°)$, to then find the length of the spiderweb.

We have

$$\sin(45° - 30°) = \sin(45°)\cos(30°) - \cos(45°)\sin(30°)$$
$$= \frac{\sqrt{2}}{2}\frac{\sqrt{3}}{2} - \frac{\sqrt{2}}{2}\frac{1}{2}$$
$$= \frac{\sqrt{6} - \sqrt{2}}{4}.$$

Let $s$ denote the length of the spiderweb, then $\sin(15°) = \dfrac{8}{s}$, so

$$s = \frac{8}{\dfrac{\sqrt{6} - \sqrt{2}}{4}} = \frac{32}{\sqrt{6} - \sqrt{2}} = 8(\sqrt{6} + \sqrt{2}).$$

Thus $A \times B = 12$.

**Answer:** 12

## Problem 20 Solution

The curve $x^2 + 4y^2 - 16 = 0$ is equivalent to $\dfrac{x^2}{4^2} + \dfrac{y^2}{2^2} = 1$, which is a horizontal ellipse centered at the origin with major axis 8 and minor axis 4.

The curve $x^2 + y^2 + 2y = 0$ is equivalent to $x^2 + (y+1)^2 = 1$ after completing the square, which is a circle centered at $(0, -1)$ and radius 1.

These are graphed on the same coordinate plane below:

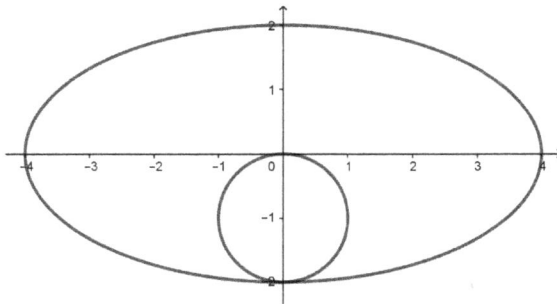

Thus, the two curves meet only at $(0, -2)$.

**Answer:** 1

# 3. Appendix

## 3.1 Division H Topics Covered

### Algebra

- Students should be comfortable with ratios, proportions, and their applications to problems involving work and motion, but these problems are not a main focus at this level
- Radicals, Exponents, and Logarithms: Simplest Radical Form for Roots, Laws of Exponents, Laws of Logarithms including change of base
- Complex Numbers: Arithmetic Operations and writing in rectangular form
- Factoring Tricks: Sums and differences of squares, cubes, etc., Expanding $(x+y)^n$ using Pascal's triangle
- Solving Equations: Linear Equations, Quadratic Equations, Systems of Equations, Substitutions to rewrite higher degree equations as quadratics, Radicals, Absolute Values
- Quadratics: Graphing and Vertex Form, Maxima and Minima, Quadratic Formula, Discriminant

- Polynomials: Polynomial Long Division, Remainder and Factor Theorem, Rational Root Theorem

## Geometry

- As a general rule students should be comfortable using algebraic techniques (linear equations, quadratic equations, systems of equations, etc.) as tools for applying the geometric concepts listed below
- Angles in Parallel Lines (corresponding angles, alternating interior/exterior angles, same-side interior/exterior angles, etc.)
- Analytic Geometry: Equations of Lines, Parabolas, and Circles, Distance Formula, Midpoint Formula, Geometric Interpretation of Slope and Angles
- Triangles: Congruence and Similarity, Pythagorean theorem, Ratios of Sides for triangles with angles of 45, 45, 90 or 30, 60, 90
- Trigonometry: General understanding of sine, cosine, tangent, and their cofunctions, Law of Sines and Cosines, Trigonometric Identities for double angles, sums/differences, etc.
- Centers in Triangles: Definitions of altitudes, medians, angle bisectors, perpendicular bisectors
- Interior and Exterior Angles of Polygons, including the sum of all these angles, each angle if the polygon is regular, etc.
- Areas and Perimeters of basic shapes such as triangles, rectangles, parallelograms, trapezoids, and circles, Heron's formula and formulas using inradius or circumradius for triangles
- Geometric Reasoning with Areas: Congruent shapes have the same area, Similar triangles have a ratio of areas that is the square of the ratio of their sides, Triangles with the same height have a ratio of their areas equal to the ratio of their bases, etc., Using multiple expressions of area to solve for unknowns

- Circles: Arc Length, Sector Area, Definitions for Tangent Lines and Tangent Circles, Inscribed Angles, Angles formed by intersecting chords
- Solid Geometry: Surface Area and Volume for Spheres, Prisms, Pyramids, and Cones, Reasoning for more general solids, such as combining the solids listed above or pieces of solids when cut by a plane, etc.

**Counting and Probability**

- Fundamental Rules: Sum and Product Rules, Permutations and Combinations
- Counting Methods: Complementary counting, Grouping objects that must be together, Inserting objects that must be apart into spaces between objects, etc.
- Sequences: Arithmetic and Geometric Sequences and Series, Finding and understanding patterns and recursive definitions for general sequences
- Probability and Sets: Definitions for event, sample space, complement, intersection, and union, Understanding the use of Venn Diagrams
- Probability: In finite sample spaces as a ratio of the number of outcomes, In geometric sample spaces as a ratio of lengths, areas, or volumes, Axioms of Probability, Independence, Conditional Probability, Law of Total Probability

**Number Theory**

- Fundamental Definitions: Prime numbers, factors/divisors, multiples, least common multiple (LCM), greatest common factor/divisor (GCF or GCD), perfect squares/cubes/etc.
- Divisibility Rules for numbers such as 2, 3, 4, 5, 8, 9, 10, 11, and how to combine the rules for numbers such as 6, 22, etc.

- (Unique) Prime Factorization and how to use the prime factorization to find the number of factors, to test whether a number is a perfect square/cube/etc, to find the LCM or GCD.
- Factoring Tricks: Factors come in pairs, perfect squares have an odd number of factors, etc.
- Remainders: Patterns for finding remainders, for example units digits or last two digits

## 3.2 Glossary of Common Mathematical Terms

**Acute Angle** An angle less than $90°$.

**Altitude of a Triangle** A line segment connecting a vertex of a triangle to the opposite side forming a right angle. Also called the height of a triangle.

**Angle** A figure formed by two rays sharing a common vertex. Often measured in degrees.

**Angle Bisector** A line dividing an angle into two equal halves.

**Arc** The curve of a circle connecting two points.

**Area** The amount of space a region takes up. Often denoted using square brackets: area of $\triangle ABC = [ABC]$.

**Arithmetic Sequence** A sequence where the difference between one term and the next is constant.

**Average** See Mean.

**Base of a Triangle** One side of a triangle, often used when the altitude is drawn from the opposite side to this base.

**Binomial Coefficient** The symbol $\binom{n}{k} = \dfrac{n!}{k!(n-k)!}$.

**Chord** A line segment connecting two points on the outside of a circle.

**Circle** A round shape consisting of points that all have the same distance (called the radius) from the center of the circle.

**Circumference** The perimeter of a circle.

**Circumscribe**  To draw a shape outside another shape so that the boundaries touch.

**Coefficient**  The number being multiplied by a variable or power of a variable. For example, the coefficient of $x^3$ in $5x^5 + 4x^3 + 2x$ is 4.

**Complement**  In probability, the complement of a set is all elements outside the set.

**Composite Number**  A number that is not prime.

**Congruent**  Two shapes or figures that are exactly the same.

**Cube**  A solid figure formed by 6 congruent squares that all meet at right angles.

**Deck of Cards**  A standard deck of cards has 52 cards. There are 4 suits (clubs, diamonds, hearts, and spades) with each suit having cards of 13 ranks ($A$ (ace), $2, 3, \ldots, 10$, $J$ (jack), $Q$ (queen), and $K$ (king) ).

**Degree of a Polynomial**  The highest power of a variable in the polynomial. For example, the degree of $2x^3 - 5x^6 + 2$ is 6.

**Denominator**  The bottom number in a fraction.

**Diagonal**  A line segment connecting two vertices of a shape or solid that is not an edge of the shape or solid.

**Diameter**  A chord passing through the center of a circle. The diameter has length that is twice the radius.

**Die or Dice**  A standard die (plural is dice) has 6 sides. Each of the 6 sides has the same chance when the die is rolled.

**Digit** One of $0, 1, 2, \ldots, 9$ used when writing a number.

**Discriminant** The expression $b^2 - 4ac$ for a quadratic equation $ax^2 + bx + c = 0$.

**Distinguishable Objects** Objects that are different.

**Divisible** A number is divisible by another number if there is no remainder when the first number is divided by the second. For example, 35 is divisible by 7.

**Divisor** A number that evenly divides another number. For example, 6 is a divisor of 48. Also called a factor.

**Edge** A line segment connecting two vertices on the outside of a shape or solid.

**Equally Likely** Having the same chance of occurring.

**Equiangular Polygon** A shape with all equal angles.

**Equilateral Polygon** A shape with all equal sides.

**Equilateral Triangle** A regular triangle, one with three equal sides and three equal angles.

**Even Number** A number divisible by 2.

**Exponent** The number another number is raised to for powers. For example, in $a$ to the power of $b$ ($a^b$), the exponent is $b$.

**Face** The shape or polygon on the outside of a solid region.

**Factor of a Number** A number that evenly divides another number. For example, 6 is a factor of 48. Also called a divisor.

**Factorial** The symbol ! where $n! = n \times (n-1) \times (n-2) \cdots \times 1$.

**Fraction** An expression of a quotient. For example, $\frac{1}{2}$ or $\frac{9}{7}$.

**Function** A function is a rule that associates exactly one output with every input. Often described using an equation.

**Geometric Sequence** A sequence where the ratio between one term and the next is constant.

**Greatest Common Divisor/Factor (GCD/GCF)** The largest number that is a divisor/factor of two or more numbers.

**Indistinguishable Objects** Objects that are the same.

**Inscribe** To draw a shape inside another shape so that the boundaries touch.

**Intersecting** Lines or curves that cross each other.

**Intersection of Two Sets** The set of objects that are in both of the two sets. Denoted using $\cap$. For example, $\{2,3\} \cap \{3,4,5\} = \{3\}$.

**Isosceles Triangle** A triangle with two equal sides and two equal angles.

**Least Common Multiple (LCM)** The smallest number that is a multiple of two or more numbers.

**Mean** The sum of the numbers in a list divided by the how many numbers occur in the list. Also called the average.

**Median** The number in the middle of a list when the list is arranged in increasing order.

**Median of a Triangle** A line connecting a vertex in a triangle to the midpoint of the opposite side.

**Midpoint**  The point in the middle of a line segment.

**Mode**  The number or numbers occurring most often in a list of numbers.

**Multiple**  A number that is an integer times another number. For example, 72 is a multiple of 8.

**Numerator**  The top number in a fraction.

**Obtuse Angle**  An angle between $90°$ and $180°$.

**Odd Number**  A number not divisible by 2.

**Parallel Lines**  Lines that do not intersect.

**Perfect Cube**  A number that is another number cubed. For example, $64 = 4^3$ is a perfect cube.

**Perfect Square**  A number that is another number squared. For example, $64 = 8^2$ is a perfect square.

**Perimeter**  The length/distance around the outside of a shape.

**Perpendicular Bisector**  A line perpendicular to and passing through the midpoint of a line segment.

**Pi ($\pi$)**  A number used often in geometry. $\pi = 3.1415926\ldots \approx 3.14 \approx \dfrac{22}{7}$.

**Polygon**  A shape formed by connected line segments.

**Polynomial**  A function that is made of adding multiples of powers of a variable. For example, $f(x) = x^4 + 3x^2 + 2x - 3$.

---

**Prime Factorization**  The expression of a number as the product of all its prime factors. For example, 24 has prime factorization $2 \times 2 \times 2 \times 3 = 2^3 \times 3$.

**Prime Number**  A number whose only factors are one and itself.

**Proportional Ratios**  Ratios that have equal values when expressed in fraction form. For example, $2 : 3$ is proportional to $8 : 12$.

**Quadratic**  A polynomial with degree 2. Often written in the form $ax^2 + bx + c$.

**Quadrilateral**  A shape with four sides.

**Quotient**  The integer quantity when dividing one number by another. For example, the quotient of $38 \div 5$ is 7 as $38 = 7 \times 5 + 3$.

**Radius of a Circle**  The distance from the center of the circle to any point on the outside of the circle.

**Randomly**  Chosen for a group of objects. Unless specified, the chance of choosing each object is the same as any other object.

**Rank of a Card**  See Deck of Cards.

**Ratio**  A relation depicting the relation between two quantities. For example $2 : 3$ or $\frac{2}{3}$ denotes that for every 3 of the second quantity there are 2 of the first quantity.

**Rational Number**  A number that can be written as a fraction.

**Reciprocal**  One divided by the number. For example, the reciprocal of 7 is $\frac{1}{7}$.

**Rectangle**  A quadrilateral with four right angles (an equiangular quadrilateral).

**Rectangular Form (of a complex number)**  A complex number written in the form $a + b \cdot i$ for real numbers $a$ and $b$.

**Regular Polygon**  A polygon with all equal sides and all equal angles (equilateral and equiangular).

**Remainder**  The quantity left over when one integer is divided by another. For example, the remainder of $38 \div 5$ is 3 as $38 = 7 \times 5 + 3$.

**Rhombus**  A quadrilateral with four equal sides (an equilateral quadrilateral).

**Right Angle**  A $90°$ angle.

**Right Triangle**  A triangle containing a right angle.

**Root of a Function**  A value of $x$ such that the function evaluates to zero. For example, $x = 2$ is a root of the function $f(x) = x^2 - 4$.

**Sample Space**  In probability, the sample space is the set of all outcomes for an experiment.

**Scalene Triangle**  A triangle with three unequal sides and three unequal angles.

**Sector**  The region formed by an arc and the two radii connecting the ends of the arc to the center of the circle.

**Sequence**  An ordered list of numbers.

**Set** An unordered collection or group of objects without repeated elements. Denoted using curly brackets. For example, $\{1,2,3,4\}$ is the set containing the integers $1,\ldots,4$.

**Similar** Shapes or solids that have the same angles and sides that share a common ratio.

**Simplest Radical Form** An expression containing a radical such that the number inside the radical is an integer that has no perfect squares.

**Sphere** A round solid consisting of points that all have the same distance (called the radius) from the center of the sphere.

**Square** A shape with four equal sides and four equal angles (a regular quadrilateral).

**Subset** A set of objects that is contained inside a larger set of objects. Denoted using $\subseteq$. For example $\{2,3\} \subseteq \{1,2,3,4\}$.

**Suit of a Card** See Deck of Cards.

**Surface Area** The total area of all the faces of a solid.

**Tangent Line** A line touching a shape or curve at exactly one point.

**Trapezoid** A quadrilateral with one pair of parallel sides.

**Triangle** A shape with three sides.

**Union of Two Sets** The set of objects that are in one or both of the two sets. Denoted using $\cup$. For example, $\{2,3\} \cup \{3,4,5\} = \{2,3,4,5\}$.

**Venn Diagram**  A diagram with circles used to understand the relationship between overlapping sets.

**Vertex**  The intersection of line segments, especially the intersection of sides or edges in a shape or solid.

**Volume**  The amount of space a solid region takes up.

**With Replacement**  When choosing objects with replacement, a chosen object is returned to the others allowing it to be chosen more than once.

## 3.3  ZIML Answers

### ZIML October 2017 Division H

| | | | |
|---|---|---|---|
| Problem 1: | 50 | Problem 11: | 67.5 |
| Problem 2: | 100.5 | Problem 12: | 313 |
| Problem 3: | 7.5 | Problem 13: | 23 |
| Problem 4: | 15 | Problem 14: | 18 |
| Problem 5: | 4.6 | Problem 15: | 18 |
| Problem 6: | 18 | Problem 16: | 4 |
| Problem 7: | 67 | Problem 17: | 5 |
| Problem 8: | 4 | Problem 18: | 12.8 |
| Problem 9: | 60 | Problem 19: | 3 |
| Problem 10: | 5 | Problem 20: | 43 |

# ZIML November 2017 Division H

| | | | |
|---|---|---|---|
| Problem 1: | 9 | Problem 11: | 5 |
| Problem 2: | 15 | Problem 12: | 1.5 |
| Problem 3: | 420 | Problem 13: | 81 |
| Problem 4: | 97 | Problem 14: | 7.1 |
| Problem 5: | 14 | Problem 15: | 8.9 |
| Problem 6: | 1 | Problem 16: | 60 |
| Problem 7: | 40 | Problem 17: | 70 |
| Problem 8: | 3.75 | Problem 18: | 66 |
| Problem 9: | 88 | Problem 19: | $-512$ |
| Problem 10: | 18 | Problem 20: | 8 |

## ZIML December 2017 Division H

| | | | |
|---|---|---|---|
| **Problem 1:** | 7 | **Problem 11:** | 19 |
| **Problem 2:** | 384 | **Problem 12:** | 7 |
| **Problem 3:** | 14 | **Problem 13:** | 7.1 |
| **Problem 4:** | 6 | **Problem 14:** | 0.44 |
| **Problem 5:** | 10 | **Problem 15:** | 8 |
| **Problem 6:** | 900 | **Problem 16:** | 25 |
| **Problem 7:** | 49 | **Problem 17:** | 72 |
| **Problem 8:** | 15 | **Problem 18:** | 25 |
| **Problem 9:** | 27 | **Problem 19:** | 25 |
| **Problem 10:** | 8 | **Problem 20:** | 28 |

# ZIML January 2018 Division H

| | | | |
|---|---|---|---|
| Problem 1: | 7680 | Problem 11: | 6 |
| Problem 2: | 23 | Problem 12: | 17 |
| Problem 3: | 0.7 | Problem 13: | 27 |
| Problem 4: | 20 | Problem 14: | 9 |
| Problem 5: | 8 | Problem 15: | $-2$ |
| Problem 6: | 32 | Problem 16: | 180 |
| Problem 7: | 8 | Problem 17: | 2 |
| Problem 8: | 64 | Problem 18: | 27 |
| Problem 9: | 180 | Problem 19: | 5 |
| Problem 10: | 3 | Problem 20: | 20 |

## ZIML February 2018 Division H

| | | | |
|---|---|---|---|
| Problem 1: | 32 | Problem 11: | 40 |
| Problem 2: | 720 | Problem 12: | 15 |
| Problem 3: | 214 | Problem 13: | 12.5 |
| Problem 4: | 5 | Problem 14: | 6 |
| Problem 5: | 15 | Problem 15: | 56 |
| Problem 6: | 197 | Problem 16: | 7 |
| Problem 7: | 76 | Problem 17: | $-15$ |
| Problem 8: | 3 | Problem 18: | 120 |
| Problem 9: | 72 | Problem 19: | 17 |
| Problem 10: | 14.9 | Problem 20: | $-26$ |

# ZIML March 2018 Division H

Problem 1:    $-17$            Problem 11:   51

Problem 2:    4                Problem 12:   56

Problem 3:    256              Problem 13:   20891898

Problem 4:    48               Problem 14:   15

Problem 5:    12               Problem 15:   27

Problem 6:    7                Problem 16:   40

Problem 7:    18               Problem 17:   462

Problem 8:    18               Problem 18:   6

Problem 9:    39               Problem 19:   14

Problem 10:   2.5              Problem 20:   13

# ZIML April 2018 Division H

| | | | |
|---|---|---|---|
| **Problem 1:** | 250 | **Problem 11:** | 91 |
| **Problem 2:** | 56.25 | **Problem 12:** | 9 |
| **Problem 3:** | 5 | **Problem 13:** | 18 |
| **Problem 4:** | 53 | **Problem 14:** | 80 |
| **Problem 5:** | 6 | **Problem 15:** | 8 |
| **Problem 6:** | 68 | **Problem 16:** | 11 |
| **Problem 7:** | 510 | **Problem 17:** | 1.4 |
| **Problem 8:** | 105 | **Problem 18:** | 6 |
| **Problem 9:** | 75 | **Problem 19:** | 720 |
| **Problem 10:** | 5 | **Problem 20:** | 6 |

# ZIML May 2018 Division H

| | | | |
|---|---|---|---|
| Problem 1: | 33 | Problem 11: | 1200 |
| Problem 2: | 56 | Problem 12: | 35 |
| Problem 3: | 19 | Problem 13: | 0.67 |
| Problem 4: | 0.9 | Problem 14: | 840 |
| Problem 5: | 24 | Problem 15: | 33 |
| Problem 6: | 56628 | Problem 16: | 13 |
| Problem 7: | 2 | Problem 17: | 5 |
| Problem 8: | $-1$ | Problem 18: | 30 |
| Problem 9: | 0.4 | Problem 19: | 6 |
| Problem 10: | $-1$ | Problem 20: | 0.33 |

# ZIML June 2018 Division H

| | | | |
|---|---|---|---|
| **Problem 1:** | 57 | **Problem 11:** | 6 |
| **Problem 2:** | 225 | **Problem 12:** | 5 |
| **Problem 3:** | 8004 | **Problem 13:** | $-16$ |
| **Problem 4:** | 5 | **Problem 14:** | 135 |
| **Problem 5:** | 10900 | **Problem 15:** | 107 |
| **Problem 6:** | 27 | **Problem 16:** | 13.3 |
| **Problem 7:** | 101 | **Problem 17:** | 6075 |
| **Problem 8:** | 12 | **Problem 18:** | 341 |
| **Problem 9:** | 829 | **Problem 19:** | 12 |
| **Problem 10:** | 7 | **Problem 20:** | 1 |

www.ingramcontent.com/pod-product-compliance
Lightning Source LLC
Chambersburg PA
CBHW072352200326
41519CB00015B/3743